AMERICAN WAR PLANS
1945–1950

OTHER BOOKS BY STEVEN T. ROSS:

European Diplomatic History, 1789–1815
From Flintlock to Rifle
The French Revolution: Conflict or Continuity?
Quest for Victory: French Military Strategy, 1792–1799

AMERICAN WAR PLANS
1945–1950

Steven T. Ross
Professor: Strategy and Policy Department,
Naval War College, Newport, Rhode Island

FRANK CASS
LONDON

First published in Great Britain and the United States of America in 1988 by
Garland Publishing Inc., New York and London

This edition first published in 1996 in Great Britain by
Frank Cass & Co. Ltd
Newbury House, 900 Eastern Avenue, London IG2 7HH

and in the United States of America by
Frank Cass
5804 N.E. Hassalo Street, Portland, Oregon 97213-3644

British Library Cataloguing in Publication Data
Ross, Steven T.
  American War Plans, 1945–50
  I. Title
  355.033573

    ISBN 0-7146-4635-0 (cloth)
    ISBN 0-7146-4192-8 (paper)

Library of Congress Cataloging-in-Publication Data
Ross, Steven T.
  American War plans, 1945–50 / Steven T. Ross.
      p. cm.
  First published: New York : Garland, 1988.
  Includes bibliographical references and index.
  ISBN 0-7146-4635-0 (cloth). -- ISBN 0-7146-4192-8 (paper)
  1. United States--Military policy. 2. United States--Foreign
relations--Soviet Union. 3. Soviet Union--Foreign relations--United
States. I. Title.
UA23.R765 1996   95-44735
355'.00973'09044--dc20   CIP

To Bea and Jay

CONTENTS

LIST OF MAPS

INTRODUCTION

The United States and the Soviet Union emerged from the Second World War as superpowers and soon became rivals. As the Cold War unfolded the possibility of armed conflict between America and Russia formed an ever present reality. The American military establishment recognized that the USSR was unlikely to launch a deliberate attack on the United States or its allies in the near future, but prudence, nevertheless, dictated the need to prepare for the remote but dangerous contingency of war against Russia. Late in 1945 planners, working for the Joint Chiefs of Staff, began to develop the first of many strategic concepts for a conflict with the Soviet Union.

The Joint Chiefs of Staff organization was in 1945 still an *ad hoc* body with no legal sanction for its existence. It was created as the result of a wartime decision to establish an Anglo-American Combined Chiefs of Staff. The JCS became the American representatives on the combined staff where they served with the British Chiefs of Staff Committee. The Joint Chiefs held their first official meeting on February 9, 1942. Although the functions of the JCS were not formally defined, the staff worked with the British on issues of Anglo-American strategic concern and advised the President on a wide range of issues including strategy and logistics. The JCS also formulated joint strategic plans and assumed responsibility for strategic conduct of military operations in areas where the United States held primary responsibility.

After V-J Day, the Joint Chiefs of Staff continued to deal with strategic issues according to its informal wartime charter until Public Law 253, the National Security Act of July 26, 1947, provided the JCS with a statutory foundation. Section 211 of the National Security Act directed the Joint Chiefs to act as the principal military advisors to the President and Secretary of Defense. The JCS was also to prepare war, mobilization, and logistics plans and provide strategic direction of the armed forces in time of war.

The World War II JCS organization included a Joint Staff which had three main components: the Joint Staff Planners, the Joint Intelligence Committee, and the Joint Logistics Committee. The National Security Act established successor committees: the Joint Strategic Plans Committee, the Joint Intelligence Committee, and the Joint Logistics Plans Committee. Another wartime committee, the Joint Strategic Survey Committee, advised the Chiefs on matters of national policy and grand strategy and continued its functions after 1947.

The Wartime Committees consisted of senior officers who supervised a number of working groups including: The Joint War Plans Committee, the Joint Intelligence Staff, and the Joint Logistics Plans Committee. The National Security Act created similar groups: the Joint Strategic Plans Group, the Joint Intelligence Group, and the Joint Logistics Plans Group. These groups usually did the basic work on the war plans. The senior committees then examined and revised them before sending them on to the Joint Chiefs. The committees could also forward the group's work directly to the Chiefs. A separate group, the State War Navy Coordinating Committee (known by the acronym SWNCC), sought to link policy and strategy in order to create coordinated initiatives. After 1947 the body was renamed the State Army Navy Air Force Coordinating Committees or SANACC. The JCS also created *ad hoc* committees for special projects.

Individual members of the JCS often followed the development of a particular plan. After officially receiving it, one or all of the service chiefs would comment on it, guided by the views expressed by their particular service staffs. The Chiefs then collectively agreed upon revisions and decided whether or not to accept a plan. If they approved, the Chiefs would then direct the Joint Strategic Plans Committee to issue implementing directives.

This volume will examine the plans for a war against the USSR. As a study in military history the book will describe how the JCS viewed the Soviet military threat and how the American military intended to counter and defeat the Russian forces. In devising strategic concepts the JCS did not operate in a vacuum. In addition to the presumed Soviet threat, a host of other factors impinged on the planning process. The Joint Chiefs had to cope with serious interservice rivalries and roles and missions controversies. They had to take into account the impact of rapid post-war demobilization upon American military readiness and capabilities. Severe fiscal constraints in defense spending after 1945 had a major influence on war planning. The Chiefs also had to consider the strategic lessons of the recent war and respond to the military implications of the new Atomic Age, topics upon which there was by no means unanimous agreement.

This book is not, however, a study of the planning process. Nor is it an examination of the bureaucratic politics of war planning. Both topics would make excellent book-length studies, but the intent of this work is to deal with the outcome of various planning efforts rather than with the process of creation. The product, not the process, is the focus of this study.

War plans and concepts of operations rarely dictate the exact course of a conflict. There is a vast difference between war on paper and real war. Enemy actions, allied views and requirements, domestic political factors and the host of operational variables that Carl von Clausewitz called friction usually combine to alter substantially a belligerent's pre-war strategy.

Prior to America's entry into the Second World War, for example, military planners decided that in case of a simultaneous conflict with Germany and Japan the United States should concentrate most of its resources against Germany and fight defensively in the Pacific until the Allies destroyed the Nazi regime. The rapid Japanese thrusts into the Pacific and Southeast Asia, however, compelled the Americans to divert large resources to the Pacific. Consequently, it was not until the end of 1943 that the United States deployed more men and material in the European Theater of Operations than in the Central and South Pacific areas. The slow build-up in Europe, in turn, played a major role in forcing the Americans to agree to British proposals for operations in the Mediterranean instead of undertaking a rapid massing of forces in the United Kingdom and launching an early invasion of France.

If war plans do not establish the precise course of a conflict, they do set the general course of strategic operations. In World War II the Americans did finally place the bulk of their military resources in the European theater, built up strength in the United Kingdom, invaded

France and drove into the heart of Germany. War Plans and concepts of operations do then set broad strategic goals and establish basic operational approaches. Had the United States gone to war with the Soviet Union during the early years of the Cold War, military operations would have followed the general outlines established in the JCS war plans, and it is these plans that form the subject of this book.

American War Plans
1945–1950

CHAPTER I

The Threat

In the years immediately following World War II the Joint Chiefs of Staff lacked a definitive expression of American policy to use as a basis for strategic planning. The Chiefs knew that U.S.–Soviet relations were deteriorating and that an emerging consensus within the political community was coming to view the USSR as an aggressive power seeking nothing less than domination of the world. Yet the same political leaders wanted a rapid demobilization of America's armed forces and drastic reductions in defense spending. Moreover, it was by no means clear how the administration intended to respond to what it perceived as a growing Soviet threat.

Political figures in the executive and congressional branches of government had a number of options under active consideration including a return to fortress America, the provision of economic aid to Europe in order to reconstitute a balance of power, diplomatic and military responses to Soviet initiatives on a case-by-case basis and the creation of regional alliance systems designed to check Russian expansion. Until the late 1940s it was not clear which policy or combination of policies would be adopted. Consequently, the Joint Chiefs had to devise their own estimates of Soviet intentions and capabilities.

In 1945 the service chiefs were men of vast experience. Generals Marshall and Arnold and Admirals King and Leahy and their successors Generals Eisenhower and Spaatz and Admiral Nimitz had guided

America's armed forces to victory in the Second World War. They understood the complexities of the higher direction of war and were equally familiar with the need to establish effective working relationships with their political leaders. Through formal and informal contacts with administration officials, the chiefs were fairly well informed about the evolving opinions of the nation's political leaders, thus mitigating to some degree the absence of clear foreign and defense policies. Consequently, in their analysis of Soviet intentions and capabilities, the chiefs' views followed the same evolutionary pattern as that of the civilian policy makers.

JCS interest in post-war relations with the USSR began even before the defeat of Germany. On May 16, 1944, the JCS informed the Secretary of State that the Soviet Union was going to emerge from the war as a leading world power. After the conclusion of hostilities, a clash between Russia and Great Britain was a possibility as the two powers moved to enhance their strategic positions, but the United States might avert such a clash by promoting continued harmony among the Big Three.[1] On August 16, 1944, the JCS again stated its belief that the USSR would emerge from the war as a major global power while England, though remaining a significant factor in world politics, would be much reduced in strength. The United States and the Soviet Union would in fact become the world's dominant powers in the wake of the Allied victory.[2]

On February 5, 1945, the JCS produced a more detailed assessment of Russian post war intentions. The Soviets would, the JCS believed, reduce their armed forces to three million active servicemen and one million recruits in order to release manpower for the massive tasks of economic reconstruction and recovery, which would not be completed before 1952. During this interval the USSR would have no economic motive for political expansion. Moscow did fear capitalist encirclement but would, nonetheless, seek accommodations with noncommunist states and attempt to avoid open clashes with the United States and Great Britain. The Soviets would try to dominate Eastern Europe and other states on their borders, and there did exist a traditional but latent desire for warm water ports. The Soviets would in more remote areas seek to prevent the formation of hostile alliance systems, but above all, Moscow would attempt to avoid hostilities with the West at least until economic recovery was complete.[3]

Later in the year a more ominous note began to appear in JCS appraisals. On October 9, 1945, the Joint Strategic Survey Committee on its own initiative reported to the Chiefs that negotiations with the USSR to establish post-war stability in Europe and the Pacific had failed to produce positive results. Moreover, Moscow, although making significant gains in the war and stretching Soviet power from the Kurile Islands to the Elbe, was still not satisfied. The Russians were currently demanding African trusteeships and putting pressure on Turkey and Norway for concessions in the Dardenelles, Spitzbergen and Bear Island. The Soviets were also carrying on large scale subversive activities in Latin America. Russian demands thrived on past success and led to new adventures. The United States for its part was rapidly reducing its military might, and once demobilized it would be virtually impossible to reconstitute rapidly American military power. To recall veterans and reconvert industry back to war production would require an extended mobilization period. The JSSC recommended that in light of Russia's aggressive attitude the United States undertake a careful examination of its military means of resistance.[4]

The JCS accepted the JSSC paper on October 15 and on the following day examined an intelligence report on Soviet military capabilities. Currently, the USSR had an army of 477 divisions. The Air Force included 350 fighter and 230 bomber regiments with a total of about 35,000 combat aircraft. With the addition of a small navy, recruits, and NKVD units the USSR had 12,700,000 men and women in uniform. Like the Americans, the Russians were quickly demobilizing, but Moscow intended to maintain substantial peacetime forces. The JCS estimated that after demobilization the USSR would have 4,411,000 troops under arms. The army would consist of 113 divisions supported by 410 air regiments and a small navy. There would also exist eighty-four satellite divisions, a figure that would grow to 114 divisions during the next three years. Although of lesser quality than Soviet formations, the satellite divisions could relieve Russian units from various occupation tasks and thus increase the number of divisions available to the Soviets for active operations.[5]

The report estimated that the Soviet Union had the military capability of overrunning Western Europe including Scandinavia and excluding Britain at any time between 1945 and 1948. The Soviets also had the forces to mount simultaneous offensives against Turkey and Iran. The Russians had only limited capabilities in the Far East, and their navy, designed primarily for coastal operations, posed no serious threat to the continental United States or to its vital sea lines of communication. The Soviet Air Force was organized for ground support of the army and for home defense; it too posed no immediate danger to the continental United States or its possessions. The Soviets were working vigorously on the development of atomic weapons, but the JCS thought that the USSR would not possess atomic bombs for at least five years. Thus, if the United States was in no immediate danger from Soviet arms, Western Europe and much of the Middle East could not effectively resist a Soviet attack. Moreover, as the Americans and British continued to demobilize the Soviet ability to overrun Europe and the Middle East would grow substantially.[6]

Despite their pessimistic view of the military balance the JCS did not in fact believe that there was any immediate danger of hostilities. The Chiefs stated that the Soviets intended to expand whenever and wherever possible but would try to avoid a major war. The lack of atomic weapons, a long-range air force, and a blue water fleet plus the need to restore the economy were factors that would lead Moscow to exercise restraint and avoid direct challenges to vital American interests.[7]

On October 23, 1945, the Joint Intelligence Staff reported on Soviet intentions and capabilities, noting that Moscow's foreign policy was imperialistic and would not change but that Russia would not in the immediate future pursue its goals by means of a general war. It would take the USSR fifteen years to restore its war losses in manpower and industry, ten years to modernize its railway and military transportation system, five to ten years to produce atomic bombs and create a long-range strategic air force, and fifteen to twenty years to build an effective ocean-going fleet.[8]

Despite structural weaknesses the USSR was, nevertheless, immensely powerful. Because of American and British demobilization and the chaotic conditions prevailing in Europe, the Soviets, even after completing their demobilization, could easily overrun the area, and by generating additional forces could also conquer Turkey and Iran. After demobilizing, the Russians could deploy thirty divisions against Western Europe, thirty against Iran and Manchuria, thirteen for home defense, and forty as an operational reserve. Satellite divisions, though not as effective as Russian units would be in their ability to keep order to rear areas, guard lines of communication and even participate in some active operations.[9] In effect the intelligence staff believed that Soviet and satellite armies could with relative ease overrun Europe and the Middle East at any time during the next several years. The Russians could not, however, threaten militarily the Western Hemisphere or Japan. Moreover, the JIS did not expect the USSR to resort to war.[10]

In the following months JCS intelligence officers continued to propound the view of Russia as a powerful nation seeking expansion by means short of war. The Joint Intelligence Staff on October 26, 1945, stated that Moscow's immediate aim was to create a security zone around the state's borders which meant continuing pressure on Turkey, Iran, Greece, Italy, and Afghanistan. The Soviet Union's long-range political objective was to gain control of the Eurasian landmass and its strategic approaches. The Soviets would, however, display great tactical flexibility and seek to avoid war with the United States.[11] On November 1, the Intelligence Staff again asserted that the USSR was trying to recover from the war and modernize the economy and would, therefore, try to avoid an armed conflict for at least a decade.[12] A report of November 8 indicated that the Russian army had about 25,000 tanks and 60 to 70,000 large caliber artillery pieces. The Russians were not going to increase their inventories but were in the process of modernizing their major weapons systems. Qualitative improvements of ground force weapons did not, however, pose a direct military threat to the continental United States.[13]

A brief report on some of America's allies, written on November 16, 1945, indicated that France was militarily quite weak and that Great Britain, acting alone, could do little more than defend the British Isles and perhaps the Cairo-Suez area.[14] Still, the Americans were not particularly alarmed for another intelligence report submitted on November 29 indicated that although the USSR could overrun Europe, Turkey, Iran and Iraq, they would not in fact launch an attack for at least five years.[15]

On January 31, 1946, the Joint Intelligence Committee again addressed the question of Soviet goals. The Committee believed that during the next five years the USSR would seek to consolidate wartime gains and obtain concessions from Turkey and Iran. Moscow would also try to expand its influence further into the Middle East.[16] The Russians would not deliberately seek a war, but the committee felt that a conflict might occur if the USSR miscalculated the willingness of other powers to resist Soviet pressure. Russia might use force against a minor power, meet opposition and compel the Western powers to assist the victim. Thus, from a local crisis a full-scale war could quickly develop.[17]

Attention shifted for a time to current problems in the Eastern Mediterranean. A Joint Intelligence Committee report of February 6, 1946, noted that British power was declining, and that if the Soviets should decide to attack into the Middle East, Great Britain alone could not halt the Russians.[18] Six days later a report noted that Soviet oil resources in the Caucasus and Rumania and industrial centers in Moscow, the Ukraine, Urals, and Silesia were within range of bombers staging out of fields in the Middle East. In case of war the Soviets would employ forty to fifty divisions to overrun the area in order to deny these air bases to the U.S. and Britain.[19]

On March 7, 1946, the Secretary of State asked the JCS to provide information on the military implications of Soviet pressure in the Middle East.[20] The JCS replied three days later. They stated that Soviet pressure on Turkey was part of a broader effort to gain a dominant position in the region, and Soviet policy constituted a direct threat to Great Britain and an indirect threat to the United States. The Chiefs continued to regard England as America's closest and most important ally and pointed out that if the Soviets successfully penetrated the Middle East, the British Empire would soon collapse, and the United States would have to face the USSR bereft of effective allies. The Chiefs did not believe that the Soviets would go to war but expected further Russian probes and pressure in the region.[21] In late May the Joint Logistics Plans Committee reported to the JCS that the Russians could logistically support simultaneous large-scale invasions of Italy and the Middle East. The Soviets could sustain a fifty-nine division force for a campaign against Italy, twenty-four divisions against Iran, twenty-five against Egypt (twenty during wet weather) and seven divisions by air against Tunisia.[22] On June 27, the Logistics committee again noted that the Russians could supply major offensive operations against Turkey, Iran, Iraq, Syria, and Egypt.[23]

The JCS also continued to examine the overall political and strategic situation. On April 6, a JCS report stated that the Soviet Union was inherently expansionist, a view that resembled George Kennan's analysis of Russian conduct. In order to preserve the peace the Chiefs believed that the United States had to demonstrate the willingness to resist aggression, if necessary, by force. In this process the Chiefs again noted their belief that Britain was America's crucial ally.[24] On April 11, the Joint Chiefs stated that in case of war the Russians could conquer the Mid-East and Western Europe. The Americans would have no strategic option but to retreat from the continent and assist the British in defending their homeland.[25]

On July 9, 1946, the Joint Intelligence Staff suggested that the Western powers might solve their problems with the USSR by negotiating a general agreement with Moscow dividing the world into spheres of influence.[26] The Chiefs never approved the proposal. On the 26th, in reply to a request from Clark Clifford for information on the overall military balance, the JCS reported that the Soviets currently had forty-two divisions in Germany supported by 4,000 combat aircraft. These forces were in a high state of readiness, could strike virtually with no warning, reach the Rhine in a day, and overrun the U.S. zone of occupation in five days. The Russians could also strike quickly in Korea, and if they did, American forces would have no choice but to pull back to Japan.[27] On the same day the Intelligence staff reported that while overrunning Europe, the Russians could also crush Turkey in a few months. They could, if they decided to attack, mobilize enough men to deploy 110 divisions against the Turks. More probably, however, the Joint Intelligence Staff expected the Soviets to continue a war of nerves against the Turks to gain political concessions from Ankara.[28]

Looking to the future, the JCS felt it had little cause to be optimistic about the military balance in Eurasia. The Joint Intelligence Staff on August 23, 1946, predicted that by 1948 the USSR would be able to deploy 2,000 bombers, 600 torpedo boats and 150 to 160 submarines against Allied lines of communication in the Mediterranean. If the Russians went to war and took Italy and Sicily, they could close the eastern Mediterranean, and if they invaded Spain, they could close the entire sea area. In either case the Allies would be faced with the prospect of mounting a major air effort to protect communication lines to Egypt and North Africa.[29]

On November 6, an intelligence estimate sought to predict the outlines of a war fought ten years in the future. By 1956 the Soviets would have a strategic air force and as many as 150 atomic weapons. The U.S. would have between 300 and 450 atomic bombs. The USSR would probably launch a conventional attack on the Eurasian landmass and withhold its atomic arsenal to deter the United States from striking the Soviet homeland. After overrunning Eurasia, the Soviets would pause in order to harness the potential of the occupied lands to their own war machine while the United States, reluctant to use atomic weapons against Western Europe, would be unable to halt the Soviet build-up.[30]

Other estimates were equally gloomy about the prospects of resisting Russian attacks. One study concluded that a Soviet invasion of Spain was logistically feasible while another indicated that after overrunning Europe the Russians could mount an aerial blitz against England with 6,500 aircraft.[31] By 1949 the intelligence analysts assumed that the Russians would have large numbers of V1- and V2-type weapons and that with conventional warheads alone had a good chance of neutralizing Britain as a military bastion.[32]

Although the American military thought that the Soviets could conquer Eurasia, they continued to believe that Moscow would not attempt to do so in the near future. On July 27, 1946, the Joint Chiefs summarized their views on Soviet intentions and capabilities. They asserted that the fundamental goal of Soviet policy was world domination, a view that was gaining wide acceptance in the civilian community. The USSR was also striving to improve its military and technological position. Russia retained the capability, even after demobilization, to conquer rapidly Europe and other contiguous areas but would probably not exercise its military option. Present Soviet policy was to consolidate recent gains, prevent the formation of an anti-Soviet bloc in Western Europe, and establish friendly regimes in Greece, Turkey, and Iran. Moscow did not desire a war in the near future, but there was always the danger that a war might arise from Soviet miscalculation.[33]

The Joint Chiefs felt that in case of war the United States would have grave problems in coping with Russia's military might. A SWNCC paper of December 1945, based in part on a JCS position paper of the previous August, stated that American military policy had as its major goals the territorial integrity of the United States and its territories, protection of the other American states and the Philippines, support of the United Nations, and the enforcement of peace terms upon the former Axis powers. World peace would henceforth be a function of continued friendly relations among the wartime Big Three. Any breakdown of cooperation required the United States to be prepared to resist aggression from its onset since America could no longer afford the luxury of initial neutrality in a major conflict. The United States, therefore, required ready well-trained forces in peacetime to hold vital strategic areas and protect the national mobilization base. America also required an extensive research and development program, a vigorous intelligence effort, and a series of detailed mobilization plans. Finally, the report noted that the United States might have to be the first power to resort to arms if it became clear that an enemy was preparing to launch an attack on America's vital interests.[34]

If war did come, the ability of the United States to strike first and even second was at best marginal. Conventional military strength declined rapidly after V-J Day. By 1946 the Army with its Air Force component declined from 8,267,000 in 1945 to 1,891,000, the Navy from 3,380,000 to 983,000, and the Marine Corps from 471,000 to 155,000.[35] On January 8, 1946, the Joint War Plans Committee issued a report providing estimates for the armed forces as of July 1946. The Committee presumed that the army would have eight divisions in Europe, all of which would be less than fifty percent effective. Six divisions in Japan and Korea would also be less than fifty percent combat ready, and the projected four division strategic reserve in the United States would have no effectiveness whatsoever.[36] Five heavy-bomber groups in Europe would be sixty-five percent effective, but five other groups in the United States constituting the reserve force would be less than twenty percent effective.[37] The Committee believed that the Marines would be in slightly better shape with two divisions seventy percent effective and a reinforced brigade ninety percent ready. The Navy, however, would have amphibious lift for only one division and a brigade.[38] The Navy would, however, be able to maintain a force of thirteen carriers at a high level of efficiency.[39] The Committee also noted that after July 1946 the strength and readiness of armed forces would continue to decline, that industry would continue converting to civilian production and that the nation lacked an effective reserve system.[40]

After 1946, demobilization and the decline of military strength continued. By 1948 the army had 554,000 troops, the Air Force 387,000, the Navy 419,000 and the Marines 84,000.[41] The Army had three divisions in the United States, four in Japan, two in Korea, and one plus three constabulary regiments in Germany. There was a single regiment in Austria and another in Trieste. Almost all of the units were engaged on occupation duties or administrative tasks, and General

Bradley later stated that in the entire army there was only one combat ready division.[42] Air Force inventory had declined from 68,400 aircraft to 20,800, about half of which were combat machines, and the navy had but 331 combat ships.[43] The draft expired in March 1947 and the government did not reinstitute a selective service system until the following year. All of the services experienced severe shortages of trained manpower.

The presence of the atomic bomb in the American arsenal was supposed to but in fact did not compensate for the weakness of the conventional force structure. Atomic weapons were in short supply. In June, 1945, the United States possessed two implosion bombs, and a year later the arsenal included nine atomic bombs, two of which were committed to tests. Furthermore, at the end of the war production of Uranium 235 and plutonium declined drastically as plants were closed, reactors shut down or run at a fraction of their capacity, and scientists, who assembled weapon cores, trigger mechanisms, and bombs, returned to civilian life. The bomb stock stood at thirteen in 1947 and grew to fifty by 1948.[44]

The bombs were mostly Mark III models that weighed 10,300 pounds and had a twenty kiloton yield. It took a specially trained thirty-nine man team two days to assemble a bomb which could remain in its ready state for only forty-eight hours after which time it had to be partially disassembled to recharge the batteries that powered the fusing and monitoring systems. There also existed a grave shortage of assembly teams. In mid-1948, for example, available teams could only put together two bombs a day, and in January, 1949, there were but seven fully trained teams.[45]

To deliver atomic weapons the Air Corps on March 21, 1946, established the Strategic Air Command. SAC's mission was to carry out long-range heavy bomber operations anywhere in the world using conventional or atomic weapons. By 1947, SAC contained six bomb groups, four of which were incomplete. Only one group, the 509th, located at Roswell Army Air Force Base in New Mexico, flew the specially modified B-29s, code-named Silver Plate, that carried atomic bombs.[46] These planes had enlarged bombbays and had to be towed over a special weapons storage pit to enable assembly teams to hoist the Mark IIIs into place. In 1946 the Air Corps had twenty-seven Silver Plates, and in 1948 there were thirty-two Silver Plates with twelve fully and eighteen partially trained crews.[47]

Crew training, however, left much to be desired. Given expected opposition, SAC intended to execute its missions at night and bomb by radar from high altitudes. Operators, therefore, had to know what scope returns to anticipate in a target area. They also needed to know nearby physical features to establish an initial point of approach. Yet crews trained mostly in daylight and operated at low altitudes to avoid equipment failure. They practiced bomb runs against radar reflectors over water or on desert ranges and did not operate in the kinds of weather they would encounter over the USSR or navigate over terrain resembling the Russian landmass. Target data, especially east of Moscow, was skimpy. In some cases the Air Corps had to use Tsarist maps for target location, and crews in any case were not assigned specific targets since SAC employed the World War II system whereby crews received specific target assignments shortly before a mission. General LeMay later said that in 1948 not a single one of his crews was able to do a professional job.[48]

The 2000 mile combat radius of the B-29 did not enable SAC to cover many Russian targets from American bases. Forward bases were not available in Western Europe. They did exist in Great Britain and Okinawa but lacked special weapons pits and atomic storage facilities. Furthermore, the Air Corps had no fighter with the range to escort the B-29s for the entire length of their missions. Aerial refueling, not introduced until 1948, improved B-29 range, and the arrival of the B-50, actually an improved B-29, and the B-36 gave promise of greater target coverage, but in 1948 the Air Force had only eighteen B-50s and four B-36s. By 1949 the Air Force had sixty-six B-29s, thirty-eight B-50s, and seventeen B-36s able to deliver atomic bombs,[49] hardly an overwhelming number.

Whether or not these aircraft in the absence of full fighter cover and complete target information could in fact have delivered their weapons in the face of fierce Soviet fighter attacks, anti-aircraft fire, and electronic countermeasures is perhaps a question that may happily remain unanswered.

In addition to material and personnel problems the military had to operate in a political vacuum when it came to dealing with the demployment of atomic weapons. For several years after the war, the government appeared to be serious about placing atomic weapons under international control. Moreover, in case of war release authority was unclear. It was not until September 16, 1948, that President Truman endorsed NSC/30 which stated that only the chief executive could order the use of atomic bombs. The document did not, however, specify the circumstances that would lead the President to order the employment of atomic weapons. In their planning efforts staff officers did not know for certain whether they could employ atomic bombs to preempt aggression, retaliate against a conventional assault, or strike back only against an enemy atomic attack.

Finally, the Joint Chiefs of Staff did not know what the strategic impact of an atomic offensive would be. On October 30, 1945, the JCS completed an analysis of the effect of atomic weapons on warfare. They approved the paper on December 8. The report stated that in the hands of an enemy atomic bombs would drastically impair American security since the United States was more vulnerable to atomic attack than the USSR. In war atomic weapons were not useful for tactical purposes, and their primary value was as a strategic weapon directed against enemy industrial and population targets. The United States required a network of bases from which to launch an atomic offensive. Since there was no defense against atomic weapons, the United States had to be prepared to strike first if it discovered that the USSR had amassed an atomic arsenal and was preparing to launch an attack. The report also noted that conventional forces were still necessary to seize and defend forward bases, deny bases to the enemy, escort bombers, and defend against enemy air attacks.[50]

The report indicated that the JCS believed that atomic bombs were important, but it made no attempt to define their precise military impact or overall effect. The example of Japan was of little use since Japan was already a defeated power by the time A-bombs were used against it, and atomic warfare had never been directed against a major power that had not yet suffered significant battlefield reverses. A Joint Intelligence Committee report of November 3, 1945 on the strategic vulnerability of the USSR to a limited air attack was also imprecise. The JIC noted that the United States lacked sufficient atomic weapons to destroy the Soviet transportation system, power grid, and metals industry. Nor, because of limited numbers, were atomic bombs useful in battlefield situations. The Air Corps could, however, attack research and development centers, aircraft and ordnance factories, and governmental administrative facilities although the absence of accurate target data meant that the successful delivery of an atomic bomb might not in fact completely destroy a designated site.[51]

The report, nevertheless, listed twenty urban centers that contained important targets. Listed in their approximate order of importance, the Committee recommended bombing Moscow, Gorki, Kuibyshev, Sverdeovak, Novosibirsk, Omsk, Saratov, Kazan, Leningrad, Baku, Tashkent, Chelyabinsk, Nizhni Tagil, Magnitogorsk, Molotov, Tbilisi, Stalinsk, Grozny, Irkutsk, and Yaroslavl. In these cities the USSR produced ninety percent of its aircraft, seventy-three percent of its guns, eighty-six percent of its tanks, forty-two percent of its steel, sixty-five percent of its refined oil and over fifty percent of its ballbearings.[52] Moreover, these cities also contained major research facilities and administrative centers, but the report cautioned that there was little data on many specific targets within particular cities, and consequently, the bombs, which lacked the yield to wipe out an entire large city, might not in fact destroy them.[53]

A month later the JIC submitted another report which was equally ambivalent about the overall effect of atomic warfare. The Committee again stated that atomic weapons were not useful tactically and could not be profitably employed against the Soviet transportation network. The most profitable targets would be production facilities and administrative centers, but because of limited numbers of weapons and the absence of accurate target data, the effects of atomic attacks might well be only temporary.[54]

The JIC did not in fact know how many atomic bombs the United States possessed. They were not until 1947 cleared to receive stockpile data, and when the Chiefs required such information, they received oral briefings. The JIC presumed that there were twenty to thirty A-bombs available. The actual figure was much smaller, but even using optimistic figures, the planners realized that submegaton weapons could not effectively destroy a target unless delivered with a high level of precision, and for many targets there was no accurate photographic or cartographic data. The report listed the same twenty cities as the November document and noted that even if attacks were successful, they would not preclude recovery in a reasonable length of time.[55]

Military leaders thus recognized that atomic weapons were enormously powerful but were still not sure of their strategic impact. The Joint War Plans Committee in the January 1946 appraisal of America's military position included a discussion of the role of atomic weapons in American strategy. The USSR, the Committee believed, was pursuing a policy of vigorous expansion. In addition to consolidating gains from the war the USSR was currently seeking additional concessions from Turkey and Iran, but Moscow needed to restore and expand the Soviet economy and modernize its military forces and would not resort to war for the next ten or fifteen years.[56]

If war did, however, occur, the United States was, because of rapid demobilization, in a weak position on the Eurasian landmass. Nor were potential allies able to offer much help. By July 1946, the British army would be reduced to twelve divisions, and the French would have a force of only eight fully equipped divisions. The Turks had a large army—forty-seven divisions—but equipment was obsolete, thus severely limiting Turkish combat effectiveness. The USSR could, therefore, overrun all of continental Europe, the Persian Gulf, Turkey, Iran, Iraq, the Levant, other regions bordering the eastern Mediterranean, Manchuria, Korea, and North China. American forces were sufficient only to protect the Western Hemisphere, primary overseas bases, Japan, India with British help, and North Africa with French and British help. The United States could not send substantial forces abroad to check the Russian advances because it would take at least nine months to prepare a two division force and its support elements for movement overseas.[57]

With the possibility of timely large-scale operations abroad eliminated from serious consideration there remained to the United States two other means of exerting military power on short notice--the Navy and the Air Corps equipped with atomic bombs. The U.S. Navy was vastly superior to the Soviet fleet, but unlike Japan in the last war, the USSR was a land power that was largely self-sufficient in raw materials. It could not be seriously impeded by American control of the sea lanes or by blockade. The Air Corps lacked the bases, personnel, and aircraft to mount large conventional air attacks, leaving atomic weapons as the only effective countermeasure available on short notice.[58]

The Committee proposed attacks against seventeen Russian cities. These cities had been designated as targets in earlier JIC studies. The original list of twenty was reduced because three cities were either out of B-29 range or of insufficient value. Within the target cities the Air Corps was to destroy industrial facilities, particularly aircraft and ordnance factories, governmental administrative complexes, and scientific research centers. Using an assumption that forty-eight percent of all A-bombs airborne would be delivered on target, the bombing scheme required ninety-eight atomic bombs plus a one hundred percent reserve for a total of 196 weapons.[59]

The report contained a number of dubious assumptions. To reach their targets the B-29s had to operate from bases in England, Foggia in Italy, Agra in India, and Chengtu in China. In 1946 none of these bases were equipped with weapons loading pits or atomic storage facilities. The report recognized that the B-29s would have to fly much of their missions without fighter escort. The B-29s would also face fanatical opposition probably including suicide attacks, but the Committee calculated the loss rate at thirty-five percent per mission. It might have been much higher. The report presumed that the B-29s would bomb at night by radar and assumed that seventy-five percent of the bombs dropped would land within 5,000 feet of the center of the target. RAF experience in night-bombing operations indicated that this calculation of percentage may have been optimistic. Moreover, given the training methods used by the 509th Bomb Group in 1946, the estimate of the CEP may also have been too favorable.

Using even the most favorable assumptions, the report was silent on the effect of atomic attacks on the Soviet ability and willingness to continue the war. Nor did the report address the question of Russian recovery capabilities. Finally, the Committee noted that it did not know how many A-bombs were available or could be ready quickly.[60] In reality the United States in 1946 had nowhere near the number of bombs projected for the air offensive, and it would be at least three years before the Americans had anything approaching the figure of 196 atomic weapons.

The JIC report thus recognized that if war ever came there was little the United States could do to halt Soviet land operations. The atomic bomb was a powerful weapon but also had limitations. Nevertheless, it was the only effective weapon the United States possessed. Because of drastic reductions in conventional forces and limited post-war defense budgets, which before 1950 never exceeded fifteen billion dollars, the JCS felt they had no choice but to rely heavily on atomic weapons.

The centrality of atomic bombs to any American response to Soviet aggression did not end discussion within the military concerning the strategic value of such weapons. On January 12, 1946, the JCS noted that weapons of mass destruction could not alone defeat the Russians. In war, conventional forces would be required not only to seize advance bases but also for driving the Soviet armies from conquered areas and possibly invading the USSR in order to conclude the conflict.[61] Such a view, of course, resembled the experience of World War II where air power paved the way for conventional land operations.

A JCS paper of March 31, 1946, put forth similar views and added that possession of an atomic arsenal plus the means of prompt delivery might in fact deter Soviet aggression.[62] War fighting rather than deterrence was, however, the dominant theme in the immediate post-war years, and on February 25, 1947, the JCS concluded that atomic bombs should be used early in any war with the USSR and called for an enlargement of the existing stockpile.[63]

Other studies placed even greater emphasis on the importance of atomic warfare. The final report of the Evaluation Board for Operations Crossroads, the A-bomb tests at Bikini, submitted on June 30, 1947, stated that an atomic blitz could not only nullify a nation's military effort but also demolish its social and economic structures for long periods of time. Moreover, in conjunction with biological and chemical warfare atomic weapons could depopulate vast areas of the earth and threaten the very existence of mankind and civilization.[64] The United States, therefore, had to amass a large number of weapons of mass destruction, and since a surprise attack with such weapons could be decisive, America had to be prepared to strike first. The report suggested that the government define the concept of incipient attack so that the United States, upon learning that another nation was preparing an atomic attack, could launch a preemptive strike.[65]

A second paper on preemption appeared on September 23, 1947. Again, the JCS urged the government to pass legislation to enable the United States to order atomic strikes when such strikes were necessary to prevent an atomic attack upon the United States.[66] The paper also called for greater efforts to learn about Soviet atomic programs and for additional studies of the impact of atomic weapons.[67]

Thus, by the end of 1947, two views concerning the strategic role of atomic weapons had emerged within the military establishment. One position, held primarily by the Army and Navy, regarded atomic weapons as a vitally important element of the nation's overall military posture. The other view, held largely by the Air Force, asserted that atomic weapons were in and of themselves decisive and could promptly destroy an enemy's will and capacity to resist. Proponents of both views agreed that atomic weapons were indeed powerful and represented the sole means of striking back against a Soviet attack. The debate over the roles and effect of atomic weapons did not, of course, end in September 1947. The military establishment continued to examine the problems associated with atomic warfare, but atomic weapons had already become a crucial element in the American arsenal. The lack of sufficient conventional forces continued to reinforce the need to rely on atomic bombs as the only effective immediate counter to armed aggression.

By mid-1946 the Joint Chiefs of Staff had come to a number of conclusions about the nature of the Soviet threat and America's ability to meet it. The JCS believed that the Soviets were aggressive, and that Moscow was not simply seeking to create buffer zones around the USSR but was intent upon world domination.

The JCS also concluded that despite the human and economic losses sustained in World War II the USSR had become a major military power. Even after demobilization, the Soviet Union maintained large standing forces bolstered by large satellite armies. American demobilization, the declining strength of the British Empire, the destruction of German and Japanese military power, the Civil War in China, and the post-war weakness of other European states served to make the Russian military position even stronger. No power could challenge the USSR on the Eurasian mainland.

The Chiefs did not, however, feel that the USSR would in the near future attempt to achieve its goals by resorting to war. Until the Soviets repaired and modernized their economy and obtained atomic weapons and a delivery system, the Chiefs presumed that the Russians would confine their efforts to expand their power and influence to means short of open armed conflict. War, if it came, would come by accident.

In such an accidental war the Chiefs assumed that the USSR could overrun Europe, the Middle East, Manchuria, North China and Korea. America lacked the conventional forces to resist the Red Army and had no allies with sufficient power to delay the Soviets until the United States mobilized. The atomic bomb was the only immediately effective weapon in the American arsenal. The Chiefs realized that a strategy based on almost exclusive reliance on atomic warfare had serious flaws, but in the absence of conventional force they had little choice but to rely heavily upon America's atomic capability.

Thus, the JCS adopted a position of military pessimism coupled with political optimism. The Soviets were in their view aggressive and powerful but would not in fact use their military power in the near future. The belief that the Soviets did not want war and that if war came it would be the result of miscalculation was not, however, entirely reassuring. The Joint Chiefs could not assume that because war was unlikely, it was, therefore, impossible. Military planners had to be ready to meet contingencies however remote, and as American-Soviet relations continued to worsen, planners within the JCS began to draw up operational concepts for a war against Russia in case the unlikely came to pass.

At the end of 1945 the planners set to work on their first concept of operations. Their initial assumptions included Soviet conventional superiority in land warfare, American superiority at sea and in atomic weapons and delivery systems, and a continuing alliance with the United Kingdom. While the JCS continued to observe the Soviet military and debated future defense organization and strategy, the Joint War Plans Committee proceeded with its tasks. In March 1946, the first draft of an operational concept for a war against the Soviet Union was ready.

NOTES

1.   James F. Schnabel. *The History of the Joint Chiefs of Staff, The Joint Chiefs of Staff and National Policy, Volume I, 1945–1947.* Wilmington, Delaware: Michael Glazier, Inc., 1979, pp. 14–15.
2.   *Ibid.*, p. 15.
3.   JCS Information memo 374, 4 February 1945.
4.   JCS 1545, 9 October 1945.
5.   JCS 203rd meeting, 16 October 1945.
6.   *Ibid.*
7.   *Ibid.*
8.   JIS 80/7, 23 October 1945.
9.   *Ibid.*
10.   *Ibid.*
11.   JIS 80/9, 26 October 1945.
12.   JIS 80/12, 1 November 1945.

13. JIS 80/14, Estimate of Soviet Postwar Military Capabilities and Intentions, 8 November 1945.
14. JIC 332, 16 November 1945.
15. JIC 250/6, 29 November 1945.
16. JIC 341, Aims and Sequence of Soviet Political and Military Moves, 31 January 1946.
17. *Ibid.*
18. JIC 342, 6 February 1946.
19. JIS 226/2, Areas Vital to the Soviet War Effort, 12 February 1946.
20. JCS 1641, 7 March 1946.
21. JCS 1441/1, 10 March 1946.
22. JCPLC, 21 May 1946.
23. JLPC, 35/16, 27 June 1946.
24. JCS 1641/4, 6 April 1946.
25. JCS 1641/5, Estimate based on assumption of occurrence of major hostilities, 11 April 1946.
26. JIS 80/26, 9 July 1946.
27. JCS to Clark Clifford, 26 July 1946.
28. JIS 253/1, 26 July 1946.
29. JIS 249/3, 23 August 1946.
30. JIS 374/1, Intelligence Estimate assuming that war between Soviet and non-Soviet powers breaks out in 1956, 6 November 1946.
31. JLPC 35/27, 21 November 1946 and JIC 375/1, 29 November 1946.
32. JIC 375/1.
33. JCS decision amending JCS 1696, 27 July 1946.
34. SWNCC 282, 19 December 1945.
35. Charles H. Donnelly. *United States Defense Policies since World War II.* Washington, D.C.: GPO, 1957. P. 78.
36. JWPC 416/1, Military Position of the United States in Light of Russian Policy, 8 January 1946.
37. *Ibid.*
38. *Ibid.*
39. *Ibid.*
40. *Ibid.*
41. Donnelly, *op. cit.*

42. *First Report of the Secretary of Defense*, Washington, D.C.: GPO, 1949, pp. 143–145; and Omar N. Bradley and Clay Blair, *A General's Life*. New York: Simon and Schuster, 1983, p. 474.

43. Alfred Goldberg, ed. *A History of the United States Air Force 1907–1957*, Princeton, N.J.: D. van Nostrand Co., 1957, p. 112; and *First Report of the Secretary of Defense*, pp. 134–135

44. David A. Rosenberg. "US Nuclear Stockpile 1945 to 1950" in *Bulletin of the Atomic Scientists*, Vol. 38, No. 5, May 1982, p. 26.

45. *Ibid.*, p. 29 and JCS 1745/15, 27 July 1948.

46. Harry R. Borowski. *A Hollow Threat: Strategic Air Power and Containment before Korea*. Westport, Conn.: Greenwood Press, 1982, p. 47. A bomber group contained thirty aircraft.

47. Rosenberg, *op. cit.*, pp. 29–30.

48. Borowski, *op. cit.*, pp. 103, 166–167.

49. *Ibid.*, 104 and Rosenberg, *op. cit.*, p. 30.

50. JCS 1477/1, Overall Effect of Atomic Bomb on Warfare and Military Organization, 30 October 1945.

51. JIC 329, Strategic Vulnerability of the U.S.S.R. to a Limited Air Attack, 3 November 1945. The target list probably came from a memo from General Norstad to General L. R. Groves in which Norstad listed sixty-six potential targets. The memo was dated September 15, 1945 and appears in an unpublished paper by David A. Rosenberg.

52. *Ibid.*

53. *Ibid.*

54. JIC 329/1, 3 December 1945.

55. *Ibid.*

56. JWPC 416/1, *op. cit.*

57. *Ibid.*

58. *Ibid.*

59. *Ibid.* The cities removed from the target list were Leningrad, Sverdlovski, and Nizhni Tagil.

60. *Ibid.*

61. JCS 1477/5, 12 January 1946.

62. JCS 147/10, 31 March 1946.

63. JCS 1745/1, 25 February 1947.

64. Final Report of the Joint Chiefs of Staff Evaluation Board for Operation Crossroads, 30 June 1947.

65. *Ibid.*
66. JCS 1805, 23 September 1947.
67. *Ibid.*

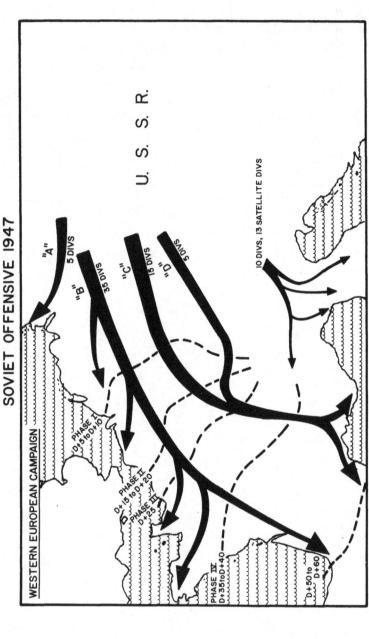

SOVIET OFFENSIVE 1947

WESTERN EUROPEAN CAMPAIGN

U. S. S. R.

"A" 5 DIVS

"B" 35 DIVS

"C" 15 DIVS

"D" 5 DIVS

10 DIVS, 13 SATELLITE DIVS

PHASE I
D+5 to D+10

PHASE II
D+15 to D+20

PHASE III
D+25

PHASE IV
D+35 to D+40

D+50 to
D+60

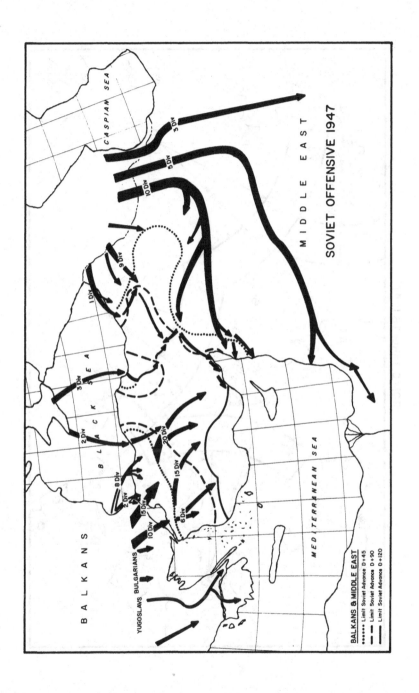

BALKANS & MIDDLE EAST

••••• Limit Soviet Advance D+45

---- Limit Soviet Advance D+90

—— Limit Soviet Advance D+120

SOVIET OFFENSIVE 1947

MIDDLE EAST

MEDITERRANEAN SEA

CASPIAN SEA

BLACK SEA

BALKANS

YUGOSLAVS

BULGARIANS

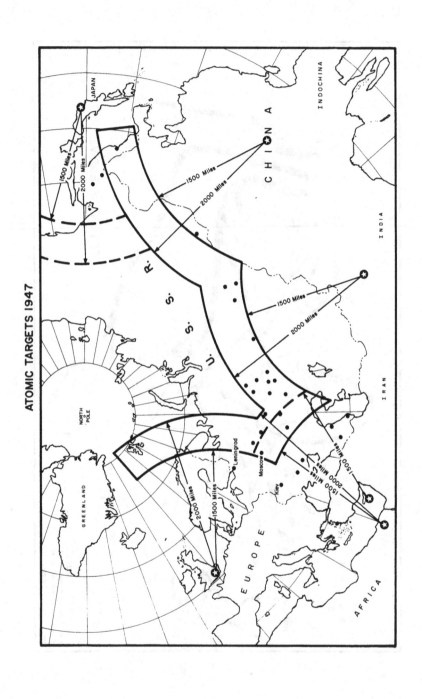

ATOMIC TARGETS 1947

CHAPTER II

The Pincher *Plans*

On March 2, 1946, the Joint War Plans Committee circulated a paper entitled "Concept of Operations for Pincher." In preparing the paper the JWPC had consulted the intelligence and logistics committees. The purpose of the paper was to establish a foundation for detailed war planning by describing in broad outline how and where a war would begin, the course of initial operations, and a strategic approach that would enable the United States and its allies to impose acceptable surrender terms upon the aggressor.

The JWPC did not believe that the USSR wanted a general war or that a conflict was likely. If war did come, it would arise from a local incident that escalated into full-scale hostilities. The planners assumed that the initial act of Soviet aggression that would spark a general war would take place in the Middle East.[1] The USSR's current goal was to establish a protective barrier of satellite countries along those portions of the Soviet border where invasion would imperil the security of vital economic and industrial regions. Moscow's objectives had been fulfilled in Eastern Europe and Manchuria, and the Soviets were currently focusing their attentions on the Middle East since domination of the region would offer protection to important Kharkov and Caucasus areas. Expansion into the Middle East would also enable the Soviets to control the Eastern Mediterranean and Persian Gulf oil resources.[2]

The Russians were in the process of putting pressure on Greece, Iran and Turkey which because of its location and large military establishment posed the most serious barrier to Soviet expansion. Domination of Turkey, therefore, was the cornerstone of the USSR's Middle Eastern policy.[3]

If unchecked, Russian expansion would threaten Britain's control of the Suez Canal and Middle East oil and would undermine England's position as a world power. The planners believed that Britain should fight if the Soviets invaded Turkey and would have to fight or accept the ultimate destruction of the empire if the Russians advanced south of Turkey. The planners believed that if the British fought while the Turks were still capable of offering effective resistance, their military situation would be vastly enhanced, but the JWPC also felt that the British would not declare war on the USSR until the Russians posed an immediate threat to the Suez Canal.[4]

War would begin when the USSR applied direct military pressure against a Middle Eastern state. The British would send limited forces to assist the victim and regional hostilities would expand into full-scale war. When hostilities began, the United States would declare a national emergency, begin to mobilize and formally enter the war six months later. For planning purposes mobilization would begin on July 1, 1947, and the U.S. would enter the war on 1 January 1948.[5]

At the start of hostilities the Red Army would consist of 113 divisions backed by eighty-four satellite units. Initially, forty Russian divisions would be available for offensive operations, a number that would grow to 100 in a short time. The Soviets would also deploy 14,350 first-line combat aircraft. British forces would number twenty divisions and 3,745 combat aircraft including the Fleet Air Arm.[6]

The Soviets would launch a two-pronged offensive to seize the upper Persian Gulf and close the Suez Canal. The Russians would succeed in their Gulf Campaign, but the JWPC presumed that the British would be able to amass sufficient strength to halt the thrust towards the Suez.[7]

Simultaneously with the initiation of their drives into the Middle East, the Russians would begin operations designed to overrun Western Europe. Moscow would not be particularly anxious to undertake the extensive operations required but would have no alternative. The Soviets would have to prevent the formation of a British-led Western European block that could constitute a continuous threat to the USSR.[8]

British, French and American occupation forces would be unable to withstand the Soviet offensive and could do little more than delay the

Red Army at the Rhine. Relatively small forces might be able to defend the Scandinavian, Danish, Italian, and Iberian peninsulas against superior numbers, but defensive operations would require a concerted effort by British, French, and American forces. The British and French governments would, however, be primarily concerned for the security of their homelands and would not divert their limited resources to Scandinavia and Denmark. Allied forces in Italy, Austria, and southern Germany might be able to establish a defense line in Italy south of the Po River, and Spain with British air and naval support and American lend-lease aid might be able to hold along the Pyrenees. On the other hand, the planners admitted that if the Soviets were willing to pay the price they could probably overrun both Italy and Spain.[9]

The staff planners did not explain how American occupation forces could fight the Red Army while the United States remained neutral. They also mentioned the possibility that the Soviets might offer immunity to U.S. forces if they withdrew peacefully from Europe. They assumed without analysis or explanation that the administration would supply Britain with shipping and lend-lease materials and would somehow join the conflict.[10]

Upon entering the war the United States would pursue operations that emphasized American superiority in air, sea, and amphibious operations and avoided a conflict of attrition against the numerically superior Russian land forces. The concept of operations required protection of the United States and its overseas territories, a relatively easy task given the absence of an effective Soviet fleet and long-range air force. American forces would next establish and defend bases in the British Isles, Egypt, India and possibly Italy and Western China. From these bases American and Allied air power would strike at the war-making capacity of the USSR while Allied fleets blockaded Russia and destroyed the Soviet war and merchant fleets. Finally, Allied forces would mount combined operations to seize selected Russian vital areas.[11]

The planners discussed various avenues of approach to these regions. They noted that because of Allied amphibious capability, operations to regain Norway and Sweden were feasible. Subsequent offensives into Northern Russia would, however, be prohibitively expensive. Allied forces would have to operate over hazardous lines of communication in areas remote from the most important Soviet economic zones, and the Red Army could amass large forces to counter any Allied expedition operating from Murmansk, Archangel, Riga, or Leningrad.[12]

An assault on Western Europe would constitute a direct frontal attack on the enemy's strongest positions. The scene of operations would be far from any Soviet vital areas, and the Allies would have to fight a war of attrition against a numerically superior foe. The planners, therefore, rejected a second Overlord.[13]

Use of a line of operations through the Mediterranean offered better prospects. The line of communications was long, and if the Soviets occupied Spain and Italy, the Allies would have to use the Red Sea route to supply forces in the Suez area and divert substantial resources to reopen the Mediterranean. On the other hand, by operating from the eastern Mediterranean, Allied forces could strike into the Black Sea and the Caucasus. Operations through the Persian Gulf and Iran would also have to operate over long communications lines, but such operations would offer useful support to forces using the Mediterranean line of approach. Therefore, the planners called for a two-pronged offensive to seize the Dardanelles and the Caucasus in preparation for additional thrusts against Soviet vital regions.[14]

The "Pincher" concept left a number of important issues unanswered. Lacking political guidance, the planners were unable to define what they meant by acceptable surrender terms. Nor were they able to describe American political objectives. The description of how war would begin was vague as was the analysis of America's role during the first six months of hostilities. It was also unclear whether or not the United States would employ atomic bombs. The planners noted that only a limited number of atomic weapons would be available when the United States entered the war. They also stated that the U.S. should exploit its superiority in scientific warfare, but nowhere did the JWPC explicitly state that the military would use atomic bombs in the offensive against Soviet war industries.

The report, despite its shortcomings, did establish a number of basic strategic assumptions that were to be the basis of all future plans for the next several years. Pincher did not describe the war beyond its initial phases but did assert that any war with the USSR, no matter how it started, would become a total global conflict. The United States continued to regard the preservation of Great Britain's security as crucial. England was America's most important ally. The planners regarded Western Europe as indefensible. If the Russians attacked,

Anglo-American troops would withdraw to the British Isles and Italy since they had no realistic expectations of halting the Red Army in Germany or the Low Countries. Moreover, the Allies would not attempt to liberate Europe by direct military action. The defeat of the Soviet Union in other areas and Moscow's final capitulation would produce the withdrawal of the Red Armies. Finally, Pincher defined the Middle East as the crucial strategic arena of World War III. Allied forces would launch part of their aerial blitz from bases in the Cairo-Suez area, and amphibious forces would attack Southern Russia via the Mediterranean and Persian Gulf.

On April 13, 1946, the Joint Staff Planners and JWPC issued another study in an attempt to clarify a number of issues raised by the report of March 2.[15] The Committee dealt at length with the role of American occupation forces and proposed that if the United States did not immediately participate in an Anglo-Soviet war, agreements should be sought with the belligerents to provide for the peaceful withdrawal of American forces. The more likely possibility, however, was that the United States would enter the war at the same time or shortly after Britain commenced hostilities. In that case American forces should either retreat from the Continent or withdraw to defensive positions in Italy or Spain. In the event that it was necessary to leave the Continent the forces should redeploy to North Africa to assist in the defense of Cairo and the Suez area. The Committee also noted that the Soviets would probably limit their operations in the Far East to the occupation of Manchuria and South Korea. American forces on the Asian mainland were vastly outnumbered, and if the Soviets attacked, they should withdraw to Japan.[16]

The planners also turned their attention to problems associated with the air offensive against the USSR. They admitted that the current state of intelligence on the Soviet economic system did not permit the creation of a detailed target profile and confined themselves to providing a list of thirty cities that contained major industrial, scientific, and governmental facilities.[17] The urban targets included the twenty cities designated in JIC 329 of November 3, 1945, plus Dnepropetrovski, Stalino, Khabarovsk, Vladivostok, Ufa, Chkalov, Kirov, Kemerovo, Komsomolsk and Zlatoust. To strike these cities the Air Corps needed bases within range of the B-29. The bases had to be defensible and capable of being supplied. The planners, therefore, chose the British Isles, Cairo-Suez, Lahore-Karachi and Chengtu as primary heavy bomber bases. Italy, Cyprus, and Crete could provide additional base areas, but their use would depend upon whether or not the Allies could prevent the Russians from overrunning these regions. In addition, other bases were required as staging areas and defensive positions from which to protect lines of supply and communication. The planners required bases in Bermuda, the Azores, Cape Verde Island, Newfoundland, Greenland, Iceland, and Benghasi.[18]

As in the original concept of operations paper the planners did not state specifically whether or not the United States would deploy atomic bombs. The only hints were the target list itself which was derived from a report listing cities designated for an atomic attack, and a statement to the effect that the absence of detailed intelligence and the availability of atomic bombs meant that urban areas in their entirety became suitable target systems as opposed to particular sites within a target city. Detailed intelligence might dictate a lower priority to area targets, but the Committee concluded that in all probability urban areas would remain high priority targets for strategic air operations.

The last section of the study dealt with the Allies' ability to defend Cairo-Suez and the Mediterranean line of communication. The planners assumed that the British would have two divisions in the Middle East on D-Day. If by D+45 days the British could reinforce their positions with one or two divisions from India, several independent brigades, and retreating Turkish forces, they could halt a twelve division Russian drive. A further four to six divisions would, however, be necessary to prevent the Soviets from building up their forces and resuming their offensive.[19]

Allied air power could prevent the Red Air Force from interdicting the Mediterranean line of communication even if Italy and Spain were overrun. The army and navy planners had a sharp disagreement over the role of carrier aircraft. The army insisted that all essential tasks could be accomplished by land-based air, and the Navy argued that carrier-based air could not only protect the Cairo-Suez region but also carry the war to Soviet bases. Since the services were currently engaged in a major debate over defense unification and the roles and missions of individual services, the Joint Staff Planners and JWPC could not arrive at a conclusion satisfactory to all and had to present a split decision in the body of the report.[20]

They did, however, agree that the Allies had to hold the Cairo-Suez area; support Spanish resistance if the Russians invaded the Iberian peninsula, and if Spain fell, neutralize Soviet invasion forces from Moroccan bases; try to hold southern Italy and Sicily, and if unsuccessful, neutralize these regions as well as southern France from bases in North Africa.[21]

After further examination, the JWPC on April 27, 1946, submitted a Joint Basic Outline War Plan which, like the March paper, was also entitled Pincher.[22] The plan was based on the earlier study and was designed to serve as the basis for preparing a joint war plan and individual service plans.

The plan presumed that the USSR did not want a general war, but its expansionist policy in the Middle East would lead to a clash with Britain. The limited conflict would quickly expand to a full-scale Anglo-Russian war. Since the survival of England was vital to the ultimate security of the United States, Washington would either declare war immediately or enter the conflict in a relatively short time.[23]

The Soviets would launch major offensives in the Middle East and Western Europe. The British would be able to retain control of the Cairo-Suez area and protect their homeland, but the Russians would be able to overrun Western Europe, the Balkans, Turkey, Iran, Iraq, the Levant, and possibly Italy and Spain. In the Far East the Russians would probably take Manchuria and southern Korea.[24]

Upon entering the war, the American military would provide for the security of the United States and its possessions, protect the Mediterranean Sea-Suez line of communications, secure and defend bases in Britain, Egypt, India and possibly Italy and Western China and from these bases mount air operations against the war-making capacity of the USSR. American forces would also blockade the Soviet Union and seek to destroy Russia's war and merchant fleets. Finally, American and Allied forces would advance through the Middle East and Persian Gulf to seize the Dardanelles-Black Sea-Caucasus area.[25] As in earlier versions the plan did not explicitly state that the United States would deploy atomic weapons. Nor did the plan attempt to describe operations after the initial phases of the war. Surrender terms and ultimate political objectives remained undefined. On the other hand, Pincher as an operational concept did seek to provide the basis for additional detailed planning. American and Allied forces were to abandon Western Europe, assume a defensive posture in the Far East, emphasize strategic air warfare and mount a counteroffensive that moved from the Middle East into southern Russian regions.

The JWPC revised the basic outline war plan and on June 18, 1946, produced an overall strategic concept and estimate of initial operations.[26] The concept-estimate paper described both the overall strategy for the first phases of a war against Russia and many of the specific operations and campaigns that would be required. The Joint Staff Planners approved the paper as a basis for further detailed service planning.

War would arise from a local clash that escalated into a general conflict. In contrast to earlier papers the JWPC assumed that the United States would be involved from the outset of hostilities. The Soviets would launch a series of offensives in Europe, the Middle East, and North China which would have great initial success.

The planners asserted that to impose American and Allied terms upon the USSR would require total mobilization of United States and Allied manpower. The Allies would, in the war, have to exploit their superiority in naval forces, their qualitative superiority in air and ground forces, and the distinct advantage of the atomic bomb. Given Allied strengths and Soviet superiority in numbers, the planners concluded that the principal initial effort against the USSR had to consist of an air offensive effort, probably deploying atomic weapons, against the Soviet Union's war-making capacity.[27]

The planners then enumerated a series of basic undertakings, military actions that constituted a first charge against American resources. The basic undertakings, if successful, would not guarantee victory but failure to carry them out would doom the United States to defeat. The United States in conjunction with its Allies had to maintain the security and war-making capacity of the Western Hemisphere and the British Isles, secure essential ground, sea and air lines of communication, provide for the security of U.S. occupation forces, guard U.S. bases and territories outside of the Western Hemisphere and provide aid to powers opposing Russian advances in critical areas.[28]

Initial operations to achieve the basic undertakings included redeploying forces as required to defend the Western Hemisphere and British Isles, establishing bases to secure the lines of communication, evacuating forces from Europe, providing support to Italy, Spain, Turkey and British forces in Egypt, evacuating Korea and North China, and, if feasible, providing aid to China. All of these operations, of course, required specific staff studies to determine the course of operations and the resources required.[29]

The planners also called for another set of operations which might or might not be executed concurrently with the initial operations in support of the basic undertakings. These operations were designed to support the overall strategic concept and included a blockade of the USSR, destruction of the Soviet fleet and the neutralization of submarine bases. Other operations were required to establish and defend bases in England, the Cairo-Suez area and, if feasible, in India, Italy, and China and the early initiation from these bases of an air offensive. Finally, the United States and its Allies had to prepare for subsequent offensives against vital Soviet areas.[30]

According to JWPC, the industrial heart of the USSR lay in the area west of the Urals and north of the Caspian and Black Seas. An advance through Western Europe would be prohibitively expensive. The Allies would face the bulk of the Red Army in regions remote from Soviet vital areas, and even if defeated in a war of attrition in Western Europe, the Russians could fall back along their own communications lines to their own home bases and reinforcements.[31]

A southern approach offered more promise. An offensive emanating from the Mediterranean, supplemented by operations from the Persian Gulf, could threaten both the Don and Baku vital areas. Moreover, Allied forces once established in the Dardanelles-Turkey-Caucasus area could construct forward bases and prepare further offensives in southern Russia. Supporting operations in the Middle East would pose difficult logistical problems especially if the Russians overran Italy and Spain, thereby forcing the Allies to supply forces in Egypt via the Red Sea. The planners, however, felt that the Soviets would also have severe logistical problems in the Middle East and assumed that Allied maritime assets could fulfill, adequately supply, and support requirements.[32]

The June Pincher study also supplied revised order-of-battle data. In 1946 the Soviets without mobilization had sixty-seven divisions available for offensive operations. By thirty days after beginning mobilization, the USSR could field 273 war-strength divisions, and by M+60 the Russians could deploy 270 divisions in Europe, forty-two in the Middle East and forty-nine in the Far East. Currently, the Red Air Force had 14,000 combat aircraft, and the Red Navy possessed 2,000 planes. Soviet air power was designed for tactical ground support and air defense, but the Russians were expected to have limited numbers of strategic bombers at the end of the three-year period covered by the plan. The Navy was weak, and the main threat posed by the Soviet fleet lay in its submarine force consisting of eighty-seven oceangoing and sixty coastal U-boats. The JWPG also noted that the Soviets were unlikely to have atomic weapons within the next several years but did possess a well-developed biological warfare capability.[33]

The Pincher strategic concept, like earlier versions, left a number of questions unanswered. The paper made no attempt to gage the impact of atomic warfare on the USSR and did not deal with operations beyond the initial Allied counteroffensive. The ability of the Allies to sustain large forces in the Middle East and Persian Gulf via the Red Sea was not explored in any detail. Nevertheless, the Joint Strategic Planners on July 8, 1946, accepted the basic concept of Pincher and directed the JWPC to keep the concept up-to-date in the light of changing conditions and to prepare a number of strategic studies of various world areas based on the Pincher concept.

The first regional plan was presented by the JWPC on August 5, 1946, and revised by the Joint Staff Planners on October 24. Entitled "Broadview," the study dealt with the defense of the Continental United States and American overseas bases and territories.[34] The Committee pointed out that the North American continent no longer enjoyed immunity from direct attack, and as the range and lethality of weapons increased, the continent's safety would continue to diminish. Against modern weapons a complete defense of the nation was not feasible. Any effort to establish a cordon defense would put an extraordinary drain on American resources and severely limit the effectiveness of offensive operations against the sources of enemy power.[35]

The Committee, therefore, proposed that defensive measures be confined to protecting vital elements of the national war potential. The main threats to America's industrial base during the next three to five years included sabotage and subversion. The JWPC believed that the Soviets had extensive secret networks in the U.S. which could do serious damage to economic, industrial and transportation targets. The Soviet forces could also conduct submarine and mining attacks on U.S. shipping and could launch limited air attacks against Alaska, Greenland, and northern Canada. The USSR could also mount a limited number of one-way air attacks and launch small commando raids against targets in the Continental United States. After 1950, the Russians would probably have the capabilities to send guided missiles against American targets as well as long-range aircraft armed with atomic weapons. The Soviets would also be able to seize territory in Alaska and Canada from which to launch missile and air assaults. Finally, the Russians might be able to mount airborne attacks on vital targets and establish secret bases from which to launch missile attacks against the Panama Canal and other overseas bases.[36]

The United States, to counter these capabilities, needed to establish an effective air warning system, an anti-aircraft gun system, and an adequate air defense system. The U.S. also needed a mobile ground reserve to counter commando and airborne raids, and naval assets to counter submarine offensives. Continuing interservice debates over roles and missions prevented the planners from assigning specific tasks to each service, and the JWPC had to content itself with describing a problem that posed a limited threat in the near future but would grow more serious as the USSR increased its long-range strike capabilities.[37]

Plan Gridle, presented on August 15, 1946, dealt with the defense of Turkey.[38] The Turkish army of forty-eight divisions was rugged and well trained. It was, however, lacking in modern equipment and by itself could do little more than delay a Russian invasion. The Air Force with fewer than 700 modern planes could offer only token resistance to Soviet air power.[39]

The Red Army, once mobilized, could send 110 divisions against the Turks, more than were actually required to overrun the country, without interfering with other major operations. A probable campaign against Turkey would involve a twenty-five division advance into Turkish Thrace while five divisions mounted parachute and amphibious assaults on both sides of the Bosporous. Ten divisions would drive into northeastern Anatolia, ten into southeastern Turkey; five would move through Iran to Baghdad and from there to Haifa, and three divisions would advance through Iran to seize Basra. The Soviet forces in Thrace would meanwhile seize a bridgehead in western Anatolia and bring their forces up to forty-one divisions. They would then strike into central and southern Anatolia supported by five divisions which made amphibious landings from the Black Sea. The Turks could hold out for about 120 days and would then be forced back to the Mediterranean coast.[40]

The JWPC noted that Turkey could play an important role in American strategy. From Turkey the Allies could block Russian offensives towards Cairo-Suez, and air power operating from Turkish fields could strike at the industrial heart of the USSR from the greatly reduced distances, and fighters based in Turkey could escort most strategic attacks.[41]

The United States should, therefore, help the Turks to strengthen their armed forces as soon as possible. The British already had a 200-man military mission in Turkey, and the United States should follow suit. Pre-war aid could transform Turkey into a defensible bastion, and if the Russians did attack before military assistance became fully effective, the Americans should, as soon as possible, strengthen the Turkish Air Force with ten fighter groups. Turkish air bases could then be enlarged to accommodate heavy bomber squadrons, and port facilities were adequate to supply air corps operations from Anatolia. Thus, the planners concluded that Turkey was strategically important and that with prompt action before hostilities began could offer a credible defense at minimum cost in Allied resources.[42]

The Caldron study, issued on November 2, 1946, dealt with the strategic situation in the Middle East.[43] The region held great economic and military value for both the USSR and Anglo-American Allies. Russia, the planners believed, had sufficient oil for its peacetime economy but insufficient developed resources to support a long, major war. The seizure of Middle East oil resources would, therefore, be a major Soviet priority in a conflict. Furthermore, the USSR had established buffer states along all important avenues of advance into Russia's industrial heartland except in the Caucasus-Black Sea area. The Middle East thus represented to the Soviets a vulnerable flank, and in a war Moscow would regard the Middle East as a source of danger and seek to conquer it quickly.[44]

Conversely, the loss of the area by Britain and the United States would deprive the Allies of the region's oil and cut the British Empire's lifeline. Moreover, denial of the Middle East would mean the loss of important bases from which the Western Powers could launch air strikes against Russian targets and mount ground counterattacks against the USSR.[45]

Order of battle data differed from other studies and from estimates provided by the Joint Intelligence Committee. The JWPC asserted that the Soviet Army on May 15, 1947, would consist of 208 divisions of which eighty-eight would be available for offensive operations. The Red Air Force would have available 13,100 combat aircraft. The satellite states would possess a total of 105 divisions and 3,309 aircraft. The Soviets, according to the JWPC, could, in theory, mobilize and expand their forces to 480 divisions within a month, but because of the strain on the economy and transportation system would probably confine the expansion of the ground forces to 360 divisions supported by 20,000 combat aircraft. The satellites could increase their forces to 136 divisions and 3,359 planes. Of this total the Soviets would employ eighty-five divisions and 2,900 aircraft for operations against Greece and the Middle East. Satellite divisions would participate in some of the operations.[46]

In the Middle East, including Greece, the British had four-and-a-half divisions and 376 combat aircraft. The United States had no ground forces in the region and one carrier air group in the Mediterranean. Both powers did, however, have the capability of moving additional forces into the area in an emergency.[47]

In the case of war the JWPC estimated that the most probable Soviet strategy would be to launch simultaneous offensives into Turkey, Greece, Iraq, and Iran. Satellite forces would operate against Greece, overrunning the northern part of the country by D+45 and all of the mainland by D+60. The Soviets by D+120 would have destroyed all general Turkish resistance. While campaigning in Turkey, five Russian divisions would move through Iran and into Iraq taking Baghdad by D+15 while three other divisions marched on Basra arriving on D+17. The Russians would also mount airborne attacks against oil refineries at Abadan and Kirkuk. After a pause to regroup, the Soviet forces at Baghdad would drive west and would reach the Mediterranean north of Haifa by D+40 and by D+60 would have pushed the British into southern Palestine.[48]

The planners also noted that the Soviets had the capability to send even larger forces into Iran and Iraq. If they chose to do so, they could overrun northern and central Iran in less than three weeks and send up to fifteen divisions, into northern Palestine by D+45. By D+165 the Russians could deploy fourteen divisions for an attack into southern Palestine, and unless the British buildup was rapid and large the Russians would have Suez within ten days.[49]

Proposed Allied courses of action depended upon specific circumstances. If the Turks managed to hold out, Allied reinforcements could provide assistance. Failing this, the Allies should try to hold Sicily, Crete, and Cyprus and from bases on these islands launch air attacks on Soviet forces and the Russian homeland. Allied reinforcements should also concentrate in Syria and Iraq to hold or at least delay the Soviet advance towards Egypt. In the final analysis holding Cairo-Suez was an irreducible minimum strategic goal, and if necessary, the Allies should sacrifice other strategic areas in the region in order to retain control of Cairo-Suez.[50]

The planners presumed that by D+90 the Allies could concentrate fourteen divisions and about 2,000 army and naval aircraft in the eastern Mediterranean. These forces could at least hold Egypt and most of Palestine. The Navy would also be able to secure the line of communication to Egypt, and the Allies would then be able to launch a strategic air offensive against the USSR. From Cairo-Suez the Allies would also build-up forces to undertake the reconquest of the Aegean Sea and Dardanelles with a view to advancing into the Russian heartland.[51]

The JWPC was careful to note that they did not seek to imply that the Caldron study implied any belief that there was going to be a war in the next three years. There was no intelligence support for such an assumption and very little support for even a possibility of hostilities between July 1, 1946, and July 1, 1949. If war should, however, arise, the planners had to be prepared to react with the forces available. They realized that given the state of Allied forces they could not halt the Soviets everywhere and established a basic strategic objective--the retention of the Cairo-Suez area. The Allies could hold Cairo-Suez with the limited forces available to them. From bases in the area Anglo-American forces could mount an aerial offensive against the USSR and marshal forces for future offensive operations.

On December 20, 1946, the JWPC produced Cockspur, an estimate of the threat to Allied forces in Italy.[52] The Allies had relatively little capability to defend Italy. The United States had one division and some service troops, a total of 34,000 men in the peninsula. There were no American combat air units in Italy. Moreover, demobilization was steadily reducing the American military presence. The British had a division, two brigade groups and some artillery formations, a total of 70,000 troops supported by 112 fighters. The British too were demobilizing. The Italian Army contained 122,500 men plus 64,120 carabinieri. The army lacked tanks and modern equipment and had a low combat efficiency rating. The Air Force with 198 operational aircraft was ill-trained; the planes were obsolete, and morale was low.[53]

The Yugoslav Army posed the most immediate threat to Italy. A battle-hardened force of 212,000 men, the Yugoslavs could send fourteen well-equipped divisions against northern Italy supported by 642 combat aircraft. The JWPC estimated that alone the Yugoslavs could not overrun all of Italy, but Soviet formations would be available to reinforce them shortly after D-Day. If the invasion of Italy took place simultaneously with a major offensive into Western Europe, five Russian divisions which had advanced into southern Germany could enter Italy via the Brenner Pass by D+20. By D+30 an additional ten divisions would be ready to advance into the Italian peninsula. If the attack on Italy were delayed until the Soviet offensive in Western Europe had overrun Germany, the Low Countries and part of northern France, all fifteen Soviet reinforcing divisions would be available by D+30.[54]

In either case the Yugoslavs and Russians could rapidly overrun the area. If the Yugoslavian Army advanced alone, it would reach Verona and Mantua by D+10 and by D+20 would deploy along a line stretching from La Spezia to Florence to Ancona. With Russian reinforcements the attack would reach Rome by D+60 and the Straits of Messian by D+75. If the attack were launched after the Soviet drive into Western Europe, the outcome of an Italian campaign would be a little different, and Yugoslav-Soviet forces would still reach the Straits by D+75.[55]

Allied alternatives were to defend the Italian-Yugoslav border, evacuate forces from Italian west-coast ports to Sicily or retreat down the Italian boot fighting delaying actions and finally withdraw to Sicily. The JWPC regarded the first option as unfeasible. The second alternative would allow the Allies to extricate their men and equipment intact although such a rapid abandonment of the mainland would have a bad effect on the morale of the Italian people. A fighting retreat would also relinquish the peninsula but would delay the enemy advance by as much as forty-five days. On the other hand, losses sustained in the fighting retreat might render the defense of Sicily impossible. The planners, therefore, concluded that the most effective Allied course of action would be an immediate withdrawal from the Italian peninsula. Forces evacuated from the mainland would deploy in Sicily where they would be reinforced by four fighter groups and could probably defend the island against any Soviet airborne or amphibious attacks.[56]

The planners presumed that in Italy as in other regions, Soviet forces had great initial advantages and could make substantial advances in the first months of a conflict. The Allies had little choice but to abandon vast areas on the Eurasian mainland while relying on air and sea power to enable them to retain lodgement areas on insular positions which could then be developed into bases for air and amphibious counterattacks. Holding Sicily would also assist the Allies in retaining control of the Mediterranean line of communication, thus giving added protection to the vital Cairo-Suez base area.

The JWPC produced an equally gloomy assessment of Anglo-American abilities to defend continental Europe on May 15, 1947.[57] The planners believed that the Western European nations had neither the political will nor the military strength to resist a Soviet attack.

The JWPC provided estimates of European nations' reactions to an invasion. East Germany would be pro-Soviet, and other satellite states would reluctantly follow Moscow's policy. Austria, Czechoslovakia, and Finland would not resist a Russian attack. France, Poland, and West Germany would react according to circumstances prevailing at the time of invasion, and it was not possible to predict their policy in advance. Only Belgium, the Netherlands, Luxemburg, Denmark, and Norway would resist. Switzerland and Sweden would remain neutral but if attacked would fight.[58]

The nations that would resist could muster only ten divisions. France, if not remaining neutral or falling into civil war, had six fully equipped divisions. The United States had one division and ten police regiments in Germany. All American formations were in a low state of combat readiness. The British had about four divisions in Germany, but demobilization was steadily reducing their strength and readiness.[59]

The Soviets on D-Day would be able to deploy one hundred divisions for a Western European campaign supported by 7,000 attack aircraft. Seven divisions would attack Norway, and three would overrun Denmark. Thirty-five divisions would advance across the North German plain and into the Low countries and northern France. Twenty divisions would move through Thuringia and the Lorraine Gap and then move down the Rhone Valley. Thirty-five divisions would remain in operational reserve.[60]

The Soviet rate of advance would be rapid. Denmark would fall by D+10, and Russian spearheads would be on the Seine by D+25. The Russians would then move air assets forward and by D+60 commence air attacks against the United Kingdom. The planners believed that the

Soviet Air Force could launch 1,400 sorties per day supplemented by eighty V-l and twenty-five V-2 missile attacks daily.[61]

The JWPC believed that the Allies had but one course of action-- withdraw from Europe, defend the British Isles, protect the line of communications to Great Britain and launch an air offensive against the USSR from British bases. Until 1948 the British could effectively handle their own defense. As the Soviets developed their arsenal of missiles, however, England's ability to defend itself would decline, and the United States would have to assist in Britain's air defense.[62]

The planners understood the value of Europe in terms of human and economic potential but believed that the current strength of European defenses, including American and British forces, was inadequate to halt a Soviet offensive. Given the current balance of forces, the JWPC had little choice but to call for the abandonment of the continent and the development of a defensive bastion and offensive base area on an insular position that could be protected against overwhelming Soviet land power.

Drumbeat, issued on August 4, 1947, dealt with the defense of the Iberian peninsula.[63] The JWPC believed that even in the event of a general war the USSR would not attack Spain. The logistics problem would be formidable; aircraft would have to be diverted from the air assault against Great Britain and the closure of the Mediterranean did not necessarily have to involve the seizure of Gibraltar since the Russians could achieve the same objective by operations from Italy.[64]

Although a Soviet invasion was improbable, the planners warned that if Moscow did decide to attack Spain the Red Army could probably overrun the Iberian peninsula unless the Allies reacted quickly. Force ratios were not favorable.

The Spanish Army contained twenty-two understrength divisions. After sixty days of mobilization, the Madrid government could bring the divisions up to full strength. In 120 days an additional 200,000 men could be mobilized bringing the army to a total of 877,000. Equipment was, however, obsolete, and overall, the quality of the Spanish Army was only fair. The Air Force was also obsolete and had but 330 operational aircraft. The Portuguese Army was poorly equipped and could not put more than two divisions in the field. The Air Force was small and its aircraft out of date. The British had a 5,000 man garrison at Gibraltar. There was no American military presence in Iberia.[65]

Should the Russians decide to undertake an invasion of Spain, they could have twenty divisions at the Pyrenees by D+45. If the French communists seized power, the Soviet Army would reach the Pyrenees ten days sooner. The Russians would then send newly mobilized divisions to reinforce the forces on the Spanish border, and by D+90 the Red Army would have fifty divisions, 800,000 men, ready for a full-scale offensive. About 1,000 aircraft would support the Army although this would force the Soviets to reduce the force designated to attack England from 6,500 to 6,000 aircraft. By D+120, however, 500 aircraft could be released from the Iberian campaign. In the absence of major Allied attacks on the Russian communication lines the build-up on the Spanish border was logistically feasible. To sustain the air campaign against Spain, the Russians would require 66,650 short tons per day, and the capacity of road, rail and water systems was about 70,000 short tons.[66]

The Soviets would attack both ends of the Pyrenees and break Spanish resistance in about twenty days. Further Spanish resistance in the absence of substantial Anglo-American support would quickly collapse, and the rest of the campaign would be largely a pursuit. In another forty days the Red Army would reach Gibraltar.[67]

The Allies had a number of options. The JWPC recommended that the United States furnish economic aid to Spain as soon as feasible in order to strengthen Madrid's capacity for military resistance. If the Soviets did decide to attack Spain, Allied air forces should attack Russian supply dumps and lines of communication. The Allies could also reinforce the Spanish forces defending the Pyrenees. The JWPC estimated that it would take thirty-four divisions to hold the Pyrenees of which the Spanish could supply twenty-two. Therefore, twelve British or American divisions supported by nine aircraft carriers and 890 landbased aircraft were required to defend the Spanish frontier. In the event that the Allies were unable to reinforce the Pyrenees in time the alternative was to defend the southern portion of the Iberian peninsula, a task requiring nine British or American divisions, fourteen Spanish divisions, nine carriers, and 739 landbased planes.[68]

If Spain insisted on neutrality in the hope of avoiding a Russian attack, the Allies had a number of options. They could assemble forces allocated to reinforce Spain in Spanish Morocco, taking the risk of local resistance. They could concentrate forces in French Morocco although they would have to fight their way ashore if the local authorities remained loyal to the government of the Russian-occupied metropole. The Allies could also concentrate forces in the Azores or United Kingdom. Finally, the Allies could violate Spanish neutrality and send forces directly to Spain.[69]

The planners believed that the Allies could mount an effective defense of Spain along the Pyrenees or north of Gibraltar if sufficient forces were allocated to the mission. Since it would take the Red Army several months to overrun France and build-up forces along the Pyrenees, the Allies would have the time to mobilize the necessary forces. Protecting Spain would also serve to guard the Mediterranean line of supply to the vital Cairo-Suez area.

The plan, of course, did not take into account Spain's diplomatic isolation in the aftermath of World War II. It might have been very difficult for Washington and London to generate support for the Franco regime prior to a major Soviet offensive. Moreover, Soviet pressure on other fronts, especially in the Middle East, might compel the Allies to divert forces designated for Iberia. Nevertheless, the planners believed that operations in the Iberian peninsula represented the outer limits of Soviet military capabilities. If the Soviets chose to attack Spain, the Allies had a good chance of successful resistance.

The JWPC presented its concept of operations for the Far East, Moonrise, on August 29, 1947.[70] For both the United States and the Soviet Union the Far East was a region of secondary strategic priority. In case of war the JWPC presumed that both powers would make their major efforts in Western Eurasia. The Russians would, however, mount a number of offensives in the Far East in order to deny to the United States bases from which Allied air power might strike at the Russian industrial heartland.[71]

Allied forces in the Far East were substantial, and if they were inferior to Russian land forces, they were superior at sea and in the air.

As of July 1, 1947, the United States maintained four divisions in Japan, two in Korea, and two Marine battalions and a 1,000-man advisory group in China. There were also 19,400 British, Indian, Australian, and New Zealand troops in Japan. The Chinese Nationalist Army contained 2,668,000 men badly organized and administered, and morale was low. Only thirty-nine divisions, trained and equipped by the United States Army, had a fair combat efficiency rating. The others were regarded as having low battle capabilities.[72]

The U.S. Navy maintained three cruisers, fifteen destroyers, and numerous support vessels in Chinese and Japanese waters. The Navy could in an emergency call upon major fleet units in Hawaii and American West Coast ports.[73]

The Air Force had about 978 aircraft in Asia and Alaska. This total included six fighter groups, two heavy and two light bomb groups, and a variety of patrol and reconnaissance squadrons. British and Imperial forces supplied an additional 212 aircraft. The Chinese Nationalists had 480 operational aircraft. The planes were modern and the pilots well trained, but repair and maintenance facilities were inadequate to sustain combat operations.[74]

The Soviet Army in the Far East consisted of thirty-two divisions of which twenty would be available for offensive operations. After thirty to forty-five days of mobilization, the Soviets would have thirty-five mobile and seven static divisions, and by D+30 the Red Army would have fifty-three divisions in the Far East. Forty-five divisions plus three outer Mongolian divisions would be used in assaults against China. About 1,150,000 Chinese communist regular troops and 2,000,000 militia were available to assist the Red Army by conducting guerrilla operations against nationalist forces.[75]

The Soviet Navy had limited capabilities. Surface vessels were few in number, and many were obsolete. The main naval threat came from ninety-nine submarines and 130 torpedo boats, but the Red Navy posed little threat to the Allies outside of its own coastal waters. The Russians had about 2,200 aircraft in the Far East and by D+135 could increase the number of available aircraft to 3,000. The Russian Air Force functioned primarily in the ground support and air defense roles and had almost no long-range strategic capability.[76]

The planners noted that Soviet logistic problems in the Far East were serious since communication lines with European Russia were limited. The Soviets had, however, built up supply bases near the Manchurian border and could sustain active operations for up to thirty-five divisions for about six months without resupply.[77]

In a global war the main Soviet effort in the Far East would be directed against China. Nine divisions would launch a concentric attack on Harbin and seize the city by D+10. A division from Korea would advance on Mukden while two divisions from Port Arthur and three Mongolian divisions would advance into the Jehol area. By D+40 these forces would have overrun Manchuria and be poised to strike at Peiping.[78]

After pausing to reorganize and resupply their forces, the Red Army would resume its attacks on D+50. Peiping would fall in about ten days. After a second pause, the Soviets would begin the third phase of their attack on D+90 by advancing toward the Yellow River. Supported by Red Chinese troops, the Soviets would capture Suchow and the river line by D+150.[79]

The Joint Intelligence Staff disagreed with the JWPC estimate of Soviet capabilities and intentions. They believed that the Red Army could reach the Yellow River by D+80 to D+90 and could reach Nanking and Hankow by D+100 to D+110. Both Committees agreed that the Russians probably would not seek to advance to the Yangtze.[80]

While attacking China, the Russians would also overrun Korea. Using five divisions, they could occupy the southern half of the country by D+20. In Japan small groups of Japanese communists would undertake subversive and sabotage operations. The Soviets also had the forces available to attack and overrun Hokkaido, but because of the difficulty of holding the island in the face of an Allied naval blockade and air attacks, the JWPC did not believe that the Russians would attempt an invasion. Finally, the Russians would undertake harassing air, naval and perhaps commando raids against American facilities in Alaska.[81]

The JWPC, in dealing with an American response to a Russian offensive, noted that the area was remote from vital regions of the USSR and that it would, therefore, be pointless to contemplate counteroffensives in the Far East. Moreover, the region contained few available natural resources and was not industrially developed. In fact, any part of the area occupied by the United States would constitute an economic liability. Allied forces in the Far East should, therefore, be held to a minimum and seek to accomplish primarily defensive tasks.[82]

The planners then called for American troops to withdraw from Korea as soon as the Red Army attacked. The troops would proceed to Japan and contribute to the defense of the main Japanese Islands. American forces in China, unable to rely upon the Nationalists, should withdraw from the mainland as soon as their positions became untenable.[83]

The United States should also send military aid to the Nationalists in hopes of complicating the Russian advance. Such aid should not, however, be of such magnitude as to detract from the overall war effort.[84]

American forces in Japan plus forces withdrawn from Korea should hold Honshu, Kyushu and possibly Hokkaido. After the start of the war, the Americans should equip up to ten Japanese divisions to assist in the defense of their homeland. Alaska and the Aleutians could be defended primarily by air and naval forces. Four fighter and two heavy bomber groups, a naval task force of cruisers and destroyers, and an army division were deemed to be sufficient. [85]

The U.S. Navy and Air Force would also carry out a number of limited offensive operations. The Navy would deny sea lines of communication to the Russians and destroy their naval and merchant fleet in the Pacific. The Air Force would attack ports, military installations, and important industrial systems in Siberia and the Maritime Provinces.[86]

The plan thus called for the United States to follow a strategy of economy of force. As in pre-World War II plans Asia was considered to be a secondary theater, and American forces would fight a war that was primarily defensive. The United States would not seriously contest a foe on the Asian mainland but would establish positions on the rim of the continent, confine Soviet power to the mainland and employ limited forces to inflict as much damage as possible on the USSR without drawing Allied strength away from decisive areas.

Deerland, the final strategic study in the Pincher series, was issued on September 30, 1947.[87] It dealt with the strategic approaches to the North American continent. American naval and air superiority would, the planners believed, be adequate to protect the northeastern approaches.

The Red Air Force lacked the long-range bomber force to directly attack North America. The Soviets possessed about 100 medium bombers that could attack Bear Island, Spitsbergen, the Faeroes, Iceland, and Jan Mayen Island, and about 100 heavy bombers that could reach Greenland and the Azores from bases in occupied Norway and Spain.[88]

The Russian surface fleet posed no serious threat. The Soviets did have about 106 submarines in the Arctic and Baltic. About ten of them were the excellent German-type XXI. The Russians could build fifty U-Boats per year. Most of them would be of the latest type, and consequently, the submarine menace would grow progressively more serious. Soviet amphibious capabilities were limited as were their airborne capacities. In the absence of organized opposition Russian troops might be able to invade outlying islands but would be unable to sustain an occupation of the westernmost positions in the face of Allied counterattacks.[89]

In war the Soviets, after occupying Norway, would seize Bear Island and Spitsbergen. The Russians would then invade Jan Mayen Island and attempt an airborne assault upon Iceland to disrupt temporarily the Atlantic line of communications and force the Allies to divert forces from more vital areas. The Soviets would also undertake active submarine operations in the Atlantic and establish weather stations in northeastern Greenland.[90]

The Allies had two strategic alternatives. They could attempt to hold Scandinavia and all of the North Atlantic islands in order to secure the northeastern approaches to the United States and Canada, protect air and sea lines and project offensive operations against the Russian landmass. The other course was to occupy the North Atlantic islands, leaving Norway and Sweden to their own devices.[91]

The JWPC indicated that although the first approach was strategically preferable, the Allies lacked the forces needed to protect Scandinavia. The Allies, therefore, had to adopt the second alternative.[92]

Of the North Atlantic islands Greenland was the most critical. In Russian hands it would provide bases for attacks on North America. In Allied possession it enabled naval and air units to protect lines of communication, and if Great Britain were neutralized or invaded, Greenland could be used as a primary base from which to launch an air offensive against the USSR. Greenland thus had to be occupied at the start of a war regardless of the views of the Danish government.[93]

The JWPC also regarded Iceland as an essential air and naval base especially if England fell. At the onset of hostilities the Americans would have to occupy Iceland with or without the consent of the Icelandic government.[94]

The Azores were also important. The islands could not be used as offensive bases but were necessary to secure the Atlantic line of communications. The planners, therefore, recommended occupation as soon as war began whether or not Portugal agreed.[95]

Spitsbergen and Jan Mayen Island, though marginally useful to the Allies, would not be occupied. They were too close to Russia, and the Allies did not wish to stretch their resources at the start of hostilities.

Reconnaissance and, if necessary, neutralization were the most economical use of Allied capabilities against these areas.[96]

The JWPC stated that the occupation of Greenland, Iceland, and the Azores was feasible. Requirements included a reinforced infantry division, a regimental combat team, four antiaircraft battalions, an airborne battalion, four antisubmarine warfare squadrons, four reconnaissance squadrons, a fighter group and three fighter squadrons.[97]

The planners were thus confident that the USSR posed no immediate threat to the continental United States or Canada. The North Atlantic islands in Soviet hands could be used as stepping stones to North America, but the Russians lacked the air and naval power to occupy and hold the most important islands. The Allies could use the islands either as bases to protect the Atlantic line of communications to England and the Middle East or as the first line of defense of the Western Hemisphere. As in World War II, the planners proposed that the islands be occupied as soon as possible without undue concern for the position of the possessing states.

Although the Pincher series concepts were never officially approved by the JCS, the Joint Staff Planners on August 29, 1947, directed the JWPC to prepare a war plan based on the strategic concepts contained in the Pincher plans.[98] Future war plans were, therefore, directly influenced by the Pincher estimates and proposals.

The JWPC assumed that the USSR had no intention of launching a general war in the near future, but if war came, it would be total, global, and protracted. The Soviets would strike first and with their superior land power make great gains on the Eurasian continent. The main Soviet thrusts would come in Europe and the Middle East with secondary operations in the Far East. The United States and its Allies would also give first priority to operations in Western Eurasia and would fight defensively in the Pacific region.

The Allies would use air power as their primary weapon against the USSR. An aerial offensive probably employing atomic weapons would devastate the USSR's war industries and substantially reduce Russia's ability to sustain its war effort. A ground offensive from Middle Eastern base areas would then strike into southern Russia in order to occupy industrial regions.

The requirement for a ground offensive indicated that the JWPC was unsure of the strategic impact of atomic weapons and implicitly assumed that atomic warfare alone would not force the Russians to capitulate.

The Pincher plans sought to emphasize Allied strengths. The initial stages of the war would, therefore, be a contest of Russian land power against Allied sea and air power. Limited Allied ground forces would seek to hold easily defensible positions from which naval and air powers could operate effectively. As in World War II, the United States would during the defensive phase of the conflict mobilize its vast human and material resources in a homeland safe from direct attack.

The planners did not describe the counterattacks in detail. Their primary concern was the first year to eighteen months of hostilities. They did, however, provide a general outline of the direction of Allied offensive operations. Striking at Russia from the Middle East was a strategy designed to threaten the Soviet heartland directly yet avoid a long war of attrition that a campaign in Western Europe would entail. Thus, for the next several years the Middle East overshadowed Europe as the focal point of American strategy. Many of the air bases for the attack on Russian industry were to be located in the Cairo-Suez area, and the amphibious expeditions against the USSR were to originate in Egypt. Even as the planners were writing the last of the Pincher series other studies emphasized the American commitment to Western Eurasia and the Middle East as the focus of strategic operations.

NOTES

1. JPS 789, Concept of Operations for "Pincher." 2 March 1946, Section III.
2. *Ibid.*
3. *Ibid.*
4. *Ibid.*
5. *Ibid.*, Assumptions.
6. *Ibid.*, Annex "A" to Enclosure "B" and Annex "B" to Enclosure "B."
7. *Ibid.*, Section IV.
8. *Ibid.*
9. *Ibid.*

10. *Ibid.*
11. *Ibid.*, Section VII.
12. *Ibid.*, Section VIII.
13. *Ibid.*
14. *Ibid.*
15. JPS 789/1, Staff Studies of Certain Military Problems Deriving from "Concept of Operations for "Pincher." 13 April 1946.
16. *Ibid.*, Appendix "A."
17. *Ibid.*, Appendix "B."
18. *Ibid.*
19. *Ibid.*
20. *Ibid.*
21. *Ibid.*
22. JWPC 423/3, Joint Basic Outline War Plan. Short Title: "Pincher." 27 April 1946.
23. *Ibid.*, Enclosure "B" Section IV.
24. *Ibid.*, Appendix to Enclosure "B."
25. *Ibid,*
26. JWPC 432/7, Tentative Over-all Strategic Concept and Estimate of Initial Operations. Short Title: "Pincher." 18 June 1946.
27. *Ibid.*, Enclosure "B" no. 4.
28. *Ibid.*
29. *Ibid.*
30. *Ibid.*
31. *Ibid.*, Appendix "A" to Enclosure "B."
32. *Ibid.*
33. *Ibid.*, Annex "A" to Appendix A to Enclosure "B."
34. JWPC 458/1, Preparation of Joint War Plan "Broadview." 5 August 1946 and JPS 815, 24 October 1946.
35. *Ibid.*
36. *Ibid.*
37. *Ibid.*
38. JWPC 467/1, "Gridle": Plan to provide such support as may be practicable to oppose Soviet moves to dominate Turkey. 15 August 1946.
39. *Ibid.*
40. *Ibid.*
41. *Ibid.*
42. *Ibid.*

43. JWPC 475/1, Strategic Study of the Area between the Alps and the Himalayas. Short Title: Caldron. 2 November 1946.
44. *Ibid.*, Appendix, part III.
45. *Ibid.*
46. *Ibid.*, Annex "G" to Appendix.
47. *Ibid.*, Annex "H" to Appendix.
48. *Ibid.*, Annex "G."
49. *Ibid.*
50. *Ibid.*, Main Report.
51. *Ibid.*, Annex "H."
52. JWPC 464/1, "Cockspur," 20 December 1946.
53. *Ibid.*, Appendix.
54. *Ibid.*
55. *Ibid.*
56. *Ibid.*
57. JWPC 474/1, 15 May 1947. This study did not have a special code name.
58. *Ibid.*
59. *Ibid.*
60. *Ibid.*
61. *Ibid.*
62. *Ibid.*
63. JWPC 465/2, The Soviet Threat against the Iberian Peninsula and the Means required to meet It. Short Title: "Drumbeat." 4 August 1947.
64. *Ibid.*, V.
65. *Ibid.*, Appendix.
66. *Ibid.*
67. *Ibid.*
68. *Ibid.*
69. *Ibid.*
70. JWPC 476/2, The Soviet Threat in the Far East and the Means required to oppose It. Short Title "Moonrise." 29 August 1947.
71. *Ibid.*, II Assumptions.
72. *Ibid.*, Appendix.
73. *Ibid.*
74. *Ibid.*
75. *Ibid.*
76. *Ibid.*

77. *Ibid.*
78. *Ibid.*, V Conclusions.
79. *Ibid.*
80. *Ibid.*, Appendix.
81. *Ibid.*, Assumptions.
82. *Ibid.*
83. *Ibid.*, Appendix.
84. *Ibid.*
85. *Ibid.*
86. *Ibid.*
87. JWPC 473/1, Strategic Study of the Northeastern approaches to the North American Continent. Short Title: "Deerland." 30 September 1947.
88. *Ibid.*, Appendix.
89. *Ibid.*, V.
90. *Ibid.*
91. *Ibid.*, Appendix.
92. *Ibid.*
93. *Ibid.*
94. *Ibid.*
95. *Ibid.*
96. *Ibid.*
97. *Ibid.*
98. James F. Schnabel. *The Joint Chiefs of Staff and National Policy, Volume I, 1945–1947.* Wilmington, Delaware: Michael Glazier Inc., 1979, p. 160.

CHAPTER III

The Bomb and Broiler

In addition to devising a basic concept of operations the JCS planners also had to provide strategic guidance for military and industrial mobilization. The Pincher plans dealt only with the initial stages of a war and contained no detailed projections of requirements for a protracted global conflict. Consequently, at the end of 1946 the Joint Chiefs ordered the planners to write an outline war plan that would serve as guidance for national mobilization in 1947.[1]

The basic assumptions concerning Soviet capabilities and intentions remained essentially the same as earlier estimates. The military viewed the USSR as an aggressive power able to deploy massive conventional forces on the Eurasian landmass.

The Joint Intelligence Staff on January 14, 1947, reported that the Soviets could mobilize as many as 245 divisions supported by 15,000 combat aircraft. In case of war they would employ 120 divisions in Western Europe, eighty-five in the Balkans and Middle East, and forty in the Far East, and by D+175 the Russians could occupy Europe and all of the Middle East up to the Suez.[2] Three days later the Joint Intelligence Committee noted that Soviet peacetime deployments in Europe amounted to 1,192,000 men and over 5,000 combat aircraft.[3]

On April 18, the JIS asserted that at anytime during the next three years the Russians could overrun Europe in forty-five days and soon thereafter launch air attacks upon the British Isles in order to neutralize them and thus deny vital forward air bases to the United States Air Force.[4]

Long-term projections were equally pessimistic. A JIC report of January 8, 1947, noted that by 1956 the USSR would possess not only numerical superiority in conventional weapons, but also technical equality with the West. The USSR would also have atomic weapons and aircraft able to deliver A-bombs against the continental United States. The Russians would be able to overrun Western Europe, neutralize and possibly occupy the United Kingdom, and utilize conquered resources for their own war effort. A regeneration of Western Europe would have but a minimal impact on the balance of power. A revived Western coalition could delay the Soviet advance and afford time for demolitions of the industrial and economic infrastructure, but only the United States and the British Empire would be able to offer effective armed resistance.[5]

On April 28, 1947, the Joint Intelligence Committee stated that currently the Russians could not launch major air strikes against the U.S. but warned that the Soviets would increase their capabilities. The Committee also noted that the USSR would eventually possess atomic weapons. In a war, the Russians would probably smuggle atomic weapons to subversive elements inside the United States for employment against major industrial targets.[6] In October, the Intelligence Committee informed the Joint Chiefs that the USSR would probably have atomic bombs in 1951 or 1952. Major Soviet targets would be American atomic bomb plants and major U.S. cities.[7]

The American strategists thus presumed that the USSR enjoyed superiority in land warfare. Moreover, this superiority would increase during the next several years as the Russians made qualitative improvements in their weapons systems. Any war plan would, consequently, have to presume an initial Soviet offensive while American and British forces gathered their strength for a series of counterthrusts.

On February 13, 1947, the JPS submitted to the Joint Chiefs an outline war plan, JCS 1725/1, designed to provide estimates of the forces required by the United States for a global conflict. The plan presumed Soviet superiority on land and Allied superiority at sea and in long-range strategic air power. The plan also assumed that atomic weapons would not be used since there might in fact be an international ban on such weapons.[8]

The mobilization plan leaned heavily on the Pincher series, and the JPS assumed that the war would open with a massive Soviet offensive. The Russians would quickly overrun Western Europe to the Pyrenees and might invade Spain. The Russians would also overrun Greece, Turkey, parts of Italy and the Middle East, Korea, Manchuria, North China, and probably Hokkaido.[9]

To avoid defeat the United States would have to protect the industrial capacities of North America and secure key bases in the United Kingdom, Cairo-Suez, and North India. The Americans would also have to protect the lines of communication to these vital areas. The United States would have to commence an immediate air offensive against Russian war industries. The major initial target system would be oil production and refining. About 80% of Russia's refineries were within range of B-29s operating from England and Cairo-Suez.[10]

In the Middle East the Allies should try to help the Turks halt the Russians. If this was not feasible, Anglo-American forces should hold Cyprus and develop it as a major air base area. In any case, the Mediterranean line of communication had to be retained in order to facilitate the buildup at Cairo-Suez. From New York to Cairo was 5,313 miles via the Mediterranean while the same route via the Horn of Africa and the Red Sea was 12,085 miles. Loss of the Mediterranean would vastly complicate the logistics problem and make the concentration of Allied forces at Cairo-Suez far more difficult.[11]

Finally, the Allies had to hold Spain to protect the Mediterranean, and they also had to establish forces in the Middle East before the Russians arrived. The logical defense line for Cairo-Suez was southern Palestine. Retaining Cairo-Suez was a basic requirement. The area was needed as a base from which to mount strategic air attacks against the Russian oil industry, and Allied counterattacks would originate from Egypt.[12]

The planners believed that any war would be a protracted conflict which would fall into four phases. From D-Day to D+12 months the Allies would devote their efforts to halting Soviet offensives. They would try to hold Sicily, Cyprus, and southern Turkey and had to retain control of the British Isles, Cairo-Suez and the Mediterranean line of communications. From D+12 to D+24 months the Allies would devote their major efforts to reducing Soviet war potential by strategic bombing. They would also seek to recapture base areas in Syria and Anatolia. From D+24 to D+36 the Allies would police the destruction of Russian industry and launch offensives into the southern regions of the USSR to force Moscow to capitulate. The final phase would be the consolidation of Allied control over Russia.[13]

Force requirements were substantial. The planners assumed that on D-Day the army required thirteen divisions. By D+12 the army would need forty-five divisions and by D+24 months, eighty divisions. Ultimately, by D+36 months, the army would consist of ninety divisions, the same number deployed during the Second World War.[14] Naval requirements included nine fleet carriers on D-Day and twenty-one by the end of the first year of hostilities.[15] The Air Force needed seventy groups and fifty-six independent squadrons on D-Day, 139 groups and 113 squadrons by D+24 months and 264 groups and 141 squadrons by D+36 months.[16]

In late July 1947 the Joint War Plans Committee submitted a mobilization plan, JWPC 486/7, that presumed that the United States would use atomic weapons against Russia. As in earlier plans the JWPC assumed that a war would open with a massive Russian offensive. Europe was viewed as being indefensible, the Pacific was seen as a primarily defensive theater, and the United Kingdom and Cairo-Suez were the strategic pivots for Allied counter-blows.[17]

The JWPC assumed that the United States would possess 100-200 atomic bombs on D-Day and designed a target system to inflict both economic and psychological damage upon the USSR.

The plan called for dropping thirty-four atomic bombs on twenty-four Soviet cities. Moscow would be hit with seven bombs, Leningrad with three and Karkov and Stalingrad with two each. Baku, Gorki, Dnepropetrovsk, Prosny, Zaporoshye, Omsk, Chelyabinsk, Molotov, Ufa, Stalinsk, Nizhny Tagil, Stalino, Sverdlovsk, Novosibirsk, Kazan, Kuibyshev, Saratov, Magnitogorsk, and Chkalov would receive one bomb each. There was also to be a 100% reserve of atomic weapons in case new attacks were necessary.[18]

Destruction of these cities would inflict substantial damage on Russia's war-making industries. The JWPC asserted that the atomic offensive would eliminate 86% of airframe production, 99% of aircraft engine plants, 56% of the arms plants, 99% of tank and self-propelled gun factories and 52% of crude oil refining facilities.[19]

The use of atomic weapons would additionally kill and wound millions of Russians. It would also disrupt substantially the functioning of Soviet society. The physical and psychological damage of an atomic attack might even force the USSR to capitulate immediately.[20] Thus, for the first time, the planners presumed that industrial targets and the Soviet population were equally important targets.

The JWPC divided the war into phases. Phase I, from D-Day to D+3 months, would witness a Soviet offensive that would overrun Western Europe, Turkey, Greece, parts of the Middle East, and North China. During this period the Allies would seek to secure the United Kingdom, the Atlantic and Mediterranean lines of communication and a Middle East base area within a quadrilateral marked by Aleppo, Mosul, Basra, and Suez. From forward bases the Allies by D+45 days would begin their strategic air offensive against Russian industry, people and morale.[21]

Phase II, from D+3 months to D+12 months, would witness an intensification of the air campaign. Allied ground and naval forces would hold Spain and if they had lost Sicily, Cyprus and Crete in the initial months of hostilities, regain the Mediterranean islands and develop them as bases.[22]

The final phase after D+12 remained undefined. The planners felt that the USSR might capitulate, in which case the Allies would send forces into Russia to police the surrender. On the other hand, if the Soviets decided to fight on out of sheer desperation, the Allies would have to mount amphibious operations against the Russian homeland.[23]

Since the United States would use atomic weapons, the forces required in 486/7 were substantially fewer than those needed by 1725/1. The Army required eighteen and one-third divisions on D-Day, the same number six months later and thirty-five and one-third divisions by D+12. The Navy called for nine carriers on D-Day, thirteen on D+6 and seventeen by D+12. The Air Force needed seventy-one groups at the start of hostilities plus fifty-six independent squadrons. Twenty-one of the groups would be composed of heavy bombers. By D+6 the Air Force requirement would grow to 106 groups of which twenty-nine were heavy bombers and eighty-eight independent squadrons. By D+12 the Air Force was to consist of 168 groups including forty heavy bombardment formations and 128 squadrons.[24]

A third mobilization guidance plan, JWPC 486/8, appeared on August 18, 1947. It dealt with the problem of fighting a war if the Russians gained control of the Mediterranean line of communications. The planners assumed that in their initial offensive surge the Russians might seize Crete, Cyprus and Sicily by amphibious and airborne assaults. The Soviets would be unable to attack Spain until D+150 days by which time Allied forces would probably be strong enough to hold the line of the Pyrenees or at least a bridgehead north of Gibraltar. The Russians could, however, move sufficient aircraft and torpedo boats to the Mediterranean islands to endanger the Allied lines of communication to Cairo-Suez.[25]

As in other plans, Cairo-Suez remained a vital forward base area from which to mount conventional or atomic air campaigns against the USSR. Even if the line of communication were severed, the planners intended to use the Cairo-Suez region and supply it via the Cape of Good Hope and the Red Sea.[26]

The Cape-Red Sea route would, of course, require more transports. On D+6 the Mediterranean-Suez route required 912 ships to sustain Allied ground and air forces in Egypt. The Cape-Red Sea route required 1,042 ships. By D+24 the Mediterranean line would have to employ 2,252 transports while the Cape line would need 3,848.[27]

The planners assumed that requisite merchant shipping was available and that the Allies could hold the Cairo-Suez area and support their forces via the Red Sea. If extensive demolitions were undertaken in eastern Turkey, Iraq and Iran, the Soviet advance would be delayed and the number of Soviet divisions that the USSR could move into the Middle East would, because of logistical constraints, be substantially reduced. The Allies could probably take advantage of the results of their demolitions by moving troops into Palestine, Lebanon, western Syria, and southern Turkey. The planners estimated that these forward positions could be held by four British and four American divisions, and that they could be supplied via the Cape and Red Sea.[28]

Further offensive operations, however, required the Allies to reopen the Mediterranean. If the Russians took Sicily, the Allies would need nine divisions and fourteen air groups to recapture the island. By D+12 months, presuming the Mediterranean was secure, the Americans could have nineteen divisions, twenty heavy bomber, ten light bomber, sixteen fighter and four all-weather fighter groups in the Cairo-Suez area. Along with British forces the Americans would be in a position to begin their counteroffensive.[29]

Further examination of the mobilization plans, however, raised serious doubts about their feasibility. Establishing bases in the United

Kingdom and the Middle East in fact required significantly more forces than the United States possessed or could rapidly generate.

On September 2, 1947, the Joint Logistics Plans Committee informed the JCS that the Cape-Cairo route by D+6 months required not the 1,042 ships called for in 486/8 but 1,788 ships. Moreover, Allied forces in the Middle East would need 32,360 short tons of supplies per day, and the Red Sea ports had a capacity of only 26,400 short tons.[30] A JCS report of October 14, 1947, indicated that by D+12 months the Air Force and Navy would need 91,332 aircraft, a requirement that would call for a massive production effort.[31] The Munitions Board on November 14, 1947, informed the JCS that the goals for aircraft production and other requirements were simply unrealistic. The military could not, the Board stated, produce the equipment and supplies needed for projected manpower requirements without long-term, costly, pre-war preparations.[32]

Even manpower requirements were at best optimistic. A logistics committee report of December 3, 1947, bluntly stated that activation and training requirements were not feasible. According to the mobilization guidance plans, the Army by D+3 months needed 2,722,000 troops, the Air Force 2,139,000 and the Navy 2,350,000. Current strengths were about 500,000 army troops, 475,000 sailors, 80,000 marines and 400,000 men in the Air Force. The logistics planners noted that to reach the projected force levels of D+3 would take twenty-six months and that even D-Day levels required mobilization nine months before hostilities. In the absence of prewar mobilization, selective service would have to induct 300,000 men per month thereby placing enormous strains on training facilities and the logistics system.[33]

On the same day the JLPC informed the service chiefs that transportation assets fell far short of troop lift requirements, and it would take sixteen months to reactivate the mothballed cargo fleet. With current maritime assets the United States could move 1,760,000 men overseas by D+6 months, but the plans called for sending 3,000,000 troops abroad.[34]

The Joint Strategic Plans Group on December 3, 1947, submitted Plan Charioteer to the JCS. Charioteer was to serve as guidance for mobilization planning and aircraft procurement for a war in 1955 or 1956. It covered in detail only the first six months of hostilities.

As in earlier war and mobilization plans the JSPG presumed that the Soviets enjoyed massive conventional superiority. By M+180 days the planners presumed that Russian armed forces would include 365 infantry, seventy-five armored, ten parachute and ten artillery divisions backed by 10,000 combat aircraft.[35] The USSR would by the mid-1950s also possess a limited stockpile of atomic bombs, sufficient long-range aircraft to deliver them to the continental United States and a substantial chemical and biological warfare capability.[36]

The Soviets would be able to overrun western Europe, part of the Middle East, Manchuria, and North China in the opening phase of the war. American economic and military aid to Greece, Turkey, France and other European states would increase their capacity to resist but not sufficiently to halt the Red Army before it seized most if not all of its initial objectives. The Russians would also attack the United States with atomic weapons and mount an extensive sabotage and subversion campaign within U.S. borders.[37]

The American strategic response would be to protect the continental United States and vital offensive base areas. If base areas were not occupied before hostilities, they would be garrisoned as soon as the war began. The United States would also mount an immediate air offensive with atomic bombs. The A-bomb offensive was to commence as soon as the war began, if necessary, from American bases and expand as new forward bases were readied.[38]

The essential base areas were the U.S., Great Britain, Greenland, Iceland, Alaska, the Middle East, Pakistan, Okinawa, and Japan. Using new B-36 bomber types with a 4,000 mile combat radius, the Air Force would strike first at Russian atomic bomb storage and production sites. Political and administrative centers were the next most important target system followed by scientific centers and industrial sites.[39]

The war would evolve in phases. The first phase, from D-Day to D+6 months, would encompass the Soviet offensive and the American strategic bombing campaign. If the Russians did not collapse by D+6, the United States and its Allies would mount ground and sea assaults while continuing the atomic bombing offensive supplemented with a conventional air campaign.[40]

Once the Russians surrendered, they would be reduced to their 1939 frontiers. Occupation troops would enforce the surrender terms and ensure that no future Russian regime could or would harbor offensive designs.

The planners estimated that on D-Day the U.S. Army required ten divisions in the continental United States, two in Alaska, one in Newfoundland, one in Greenland, one and a third in Iceland, one in Hawaii, two and a third in the Caribbean, and three in the Pacific. In addition, the Army had to deploy over 200 battalions of anti-aircraft artillery. The Air Force needed ninety-one and a third groups with 4,077 aircraft. Nine groups would execute the strategic bombardment mission, and the Navy had to deploy eight fleet carriers, 2,630 aircraft, and 2,840 replacements.[41]

On the same day the JSPC provided more specific figures for aircraft requirements. The Navy required a total of 14,474 aircraft and the Air Force 20,599 for a war between 1948 and 1952. After 1952, the Navy would need 17,472 planes and the Air Force 46,599 aircraft.[42]

The Charioteer Plan emphasized the importance of the atomic bomb offensive. It became the centerpiece of American strategy, and the planners strongly implied that atomic warfare alone could be decisive in a war against Russia. Shrinking forces and shrinking budgets convinced planners that effective forward defense in Europe was not feasible. Coupled with a belief in the effectiveness of air power loudly trumpeted by the Air Force, planners had to emphasize their most effective weapon. Thus the A-bomb, despite its scarcity and severe limitations in delivery capability, took on an increasingly important role in American strategy.

Emphasis on the strategic employment of atomic weapons was reflected in Plan Broiler, a short-range emergency war plan written in conjunction with the mobilization studies. On August 29, 1947, the Joint Strategic Plans Committee directed the Joint Strategic Plans Group to develop a joint outline war plan using the strategic concepts contained in the Pincher studies. The JSPG was to assume that war had been forced on the United States during the fiscal year 1948. The overall concept of operations would be based on the forces available on March 1, 1948.

Broiler was first briefed on November 8, 1947. The plan required the early deployment of atomic weapons. The JSPG admitted that they did not know how many bombs were available. Nor did they know the production rates for additional weapons. They simply assumed that there were enough A-bombs available to mount a bombing campaign sufficiently effective to force the USSR to surrender or to reduce Russian war potential to a point where the Allies could mount successful counterattacks.[43]

The planners noted serious shortages in a number of critical areas including strategic bombers, all-weather fighters, anti-aircraft battalions, technical personnel, ground divisions, and numerous ship types. Given conventional force weakness coupled with massive Soviet strength, the JSPG noted that the strategy of establishing bases for strategic bombardment in England, Cairo-Suez, India, and Okinawa was fraught with risks. The Mediterranean and Cairo-Suez areas would be particularly hard to defend.[44] The almost exclusive reliance on atomic warfare was, therefore, not based on confidence in America's superior strength but on a feeling of weakness amounting almost to desperation. On December 18, 1947, the Joint Strategic Plans Committee instructed the Joint Strategic Plans Group to revise Broiler on the assumptions that war would commence during fiscal year 1949 and that the force levels would be those of July 1, 1948. The JSPG submitted the slightly revised plan on February 11, 1948.

The revised Broiler plan assumed that the United States would use atomic bombs at the outset of hostilities. This was, in fact, *the* major premise of the plan, for the JSPG noted that the USSR and its satellites held vast superiority in ground and tactical air forces over the Allies. The Allies in fact had no choice but to avoid committing forces to oppose the Russians except where necessary to secure bases from which atomic strikes would be launched and to rely almost exclusively on atomic bombs to produce a favorable decision.[45]

American and Allied political goals were vague. In the absence of a definitive policy statement from the civilian leadership, the planners assumed that the U.S. would seek to destroy the war-making capacity of the USSR to the extent that the Americans could compel the withdrawal of Soviet political and military forces at least to within Russian 1939 borders. It was not clear whether or not the Soviets would be required to relinquish territory taken from Poland in 1939, but the planners obviously expected the Baltic states to regain their independence, and presumably Finland and Rumania would regain territories lost in 1940. The planners also assumed that the United States would create conditions within the USSR which would assure the abandonment of Soviet political and military aggression. Whether or not this meant the overthrow of the Communist regime or merely the installation of a peacefully inclined Politburo remained unclear. Finally, the JSPG stated that the United States would establish conditions conducive to future international stability.[46] There was no further elaboration, but it was clear that in order to achieve these goals the United States and its Allies would have to impose conditions tantamount to unconditional surrender upon the USSR.

According to the JSPG, the USSR's ultimate goal was Soviet domination of a Communist world. The intermediate objective was to extend Soviet control over the Eurasian continent and its strategic approaches, and Moscow's immediate aim was to establish a barrier of Soviet-dominated states around Russia's borders. This goal had been largely achieved except in the Middle East, which, as in the Pincher studies, was viewed as the most probable flashpoint.[47]

The planners did not believe that the USSR was seeking an armed conflict with the West in the near future. Rather, the Russians sought to expand by means short of war, but the possibility remained that Moscow might miscalculate the West's willingness to resist and inadvertantly spark a general war.[48]

If war did come, the Soviets would seek to seize Middle East oil resources, destroy forces of the Western powers on the Eurasian landmass, take or neutralize areas from which the U.S. might strike at the USSR, especially the Cairo-Suez area and the United Kingdom, and expand positions in China, Manchuria, and Korea. The Soviets would launch these simultaneous offensives not only to achieve their political goals but also to extend buffer areas to gain added protection from atomic attacks.[49]

The Soviet ground forces numbered 173 divisions and the satellites could contribute an additional sixty-eight divisions and twenty-five brigades. The Russians could also mobilize over three million additional troops, and the satellite states could expand their forces to 134 divisions. The air force consisted of 13,000 combat aircraft, and the number could be expanded to 20,000 in 150 days. The satellites possessed about 750 combat planes. The USSR's navy was weak but did contain 233 submarines.[50]

America's armed forces comprised of nine divisions and nine regiments, six of which were constabulary units. The Navy had eleven carriers, the Marine Corps contained one division, and the Air Force consisted of fifty-three and two-thirds groups and fifteen squadrons. The total included one heavy bomber and twelve medium bomber groups. The British Empire could contribute eight divisions, four carriers, and about 1,000 combat aircraft.[51]

With their massive strength the Red Army would quickly overrun Western Europe. Using fifty divisions and 6,500 combat aircraft, the Russians would occupy Norway and reach the Pyrenees by D+70 days. The planners could not agree on whether or not the Soviet Union would invade Spain, but if Moscow chose to move into Iberia, the Russians would reach Gibraltar by D+180. Yugoslav forces would attack Italy backed by Soviet reinforcements. Sicily would fall to them by D+105. Satellite divisions would also overrun Greece. The campaign would last sixty to ninety days depending upon the ability of the Greeks to demolish their rail and road networks. Special Soviet units could capture the Aegean Islands and Crete by D+145.[52]

The Soviets would send thirty-two divisions against Turkey, supported by 1,050 combat aircraft. The Soviet army would reach the Bosporous by D+60 and Ankara by D+180. The Russian timetable might be accelerated if the Allies failed to provide the Turks with adequate logistic support and if the Turks failed to base their strategy on fighting delaying actions back to southwestern Anatolia. In a worst case situation, the Russians could overwhelm the Turks in ninety days and push on to capture Suez by D+175 days.[53]

If the Soviets took Suez, Sicily, and Crete they could pose a serious threat to the Mediterranean line of communications. The fall of Spain and Gibraltar would enhance Russian interdiction capabilities. In any case the Allies would have to mount a major campaign to reopen and secure the Mediterranean and such a campaign would place a substantial drain on Allied resources.[54]

In the Middle East the planners estimated that the USSR would on D+10 launch airborne operations against Basra, Kirkuk, Kerman Shah, Hamaden, and Mosul in order to seize oil refineries and wells. In addition, the Soviets would send two divisions overland to Basra via Tehran and one division to Baghdad. By D+60 the Russians would also mount a major drive on Cairo-Suez, and if Allied forces in the region on D-Day were not quickly reinforced, the Red Army would strike for the Nile Delta on D+180 days.[55]

The USSR would pursue operations in the Far East to the extent that they not interfere with campaigns in Western Eurasia. Nevertheless, the Red Army would be able to occupy Korea, Manchuria, and North China by D+150 days.[56]

The Soviets would by air assaults seek to neutralize Great Britain. They would initiate a full air offensive on D+60 days using 2,700 bombers, 2,700 fighters and 1,000 ground attack aircraft. By D+12 months the Red Air Force would be able to deploy 9,000 aircraft against England and could generate 1,550 sorties per day, all escorted by fighters. The Russians would also employ V-1- and V-2-type missiles. The JSPG believed that the British would be able to defend themselves at least to the point where the United Kingdom could be used as a base. The Joint Intelligence Staff, however, disagreed, asserting that by 1948 the Soviets would be able to neutralize the British Isles. This disagreement remained unresolved.[57]

The Soviet Union would also strike directly at the United States in an effort to reduce America's productive capacity. Lacking a blue water fleet and a long-range strategic air force, the Russians would rely primarily on subversion and sabotage.

In the United States the Communist Party had 74,000 members plus numerous sympathizers. Trusted party members had, according to planners, attended special sabotage schools. Moreover, Communists controlled at least seventeen unions, including the American Communications Union, the National Maritime Union, the Transport Workers of America, the International Fur and Leather Workers, the United Furniture Workers Union, and the Office and Professional Workers Union. The total membership was 1,206,800. Communists had also infiltrated four other unions including the UAW.[58]

Upon the outbreak of war communists could disrupt, by slowdowns, strikes, sabotage and racial disturbances, the following industries: transportation, automobile and aircraft production, electrical equipment, shipping, communications, food, warehousing, farm equipment metals, furs and leather manufacturing, tobacco, and furniture making.[59]

Sabotage efforts would reduce substantially the capabilities of railroads, atomic bomb production, munitions industries, and chemical and electrical industries. Sabotage would also disrupt the water systems of all major industrial centers. Finally, secret agents in government and industry would keep the USSR informed of the situation within the United States.[60]

Thus, for the first time JCS planners identified domestic subversives as a substantial military menace. Although it is difficult to understand how furriers and furniture makers constituted a significant threat to American national security, there is no doubt that the JSPG regarded the problem of subversion as serious and concluded that substantial forces would have to be devoted to internal security in case of war.

The planners generally assumed that the initial Soviet advantage in conventional forces could not be countered by Allied mobilization of similar forces for a substantial period. Forces in being, therefore, had to be employed on a strict austerity basis to accomplish only tasks of the highest priority.

The basic Allied strategy would be to hold or seize base areas from which to launch a strategic atomic air offensive against the USSR's industrial structure and population. If the overall strategic concept was to be implemented, certain basic undertakings had to be successfully accomplished.

The war-making capacity of the United States had to be defended. To achieve this objective an air defense network able to cope with one-way air attacks had to be in place before D-Day. Five fighter groups were required to defend the continental U.S., and two and two-thirds groups would deploy to outlying bases. To cope with sabotage and subversion, the planners decided to guard only the most essential assets, including the four factories and three assembly and storage plants required to produce atomic bombs and the Soo Canal. Their defense required twenty-five anti-aircraft battalions and the equivalent of an infantry division. A further two divisions would secure outlying bases. The civil authorities—local, state, and federal—would have to cope with all other subversive activities.[61]

The planners, with the exception of the intelligence staff, assumed that the British could protect their island kingdom at least to the point where it could function as an operational base for the U.S. Air Force. The Americans would supply lend-lease-type aid and assist in convoy escort operations. Aside from the defense of the British Isles, American forces would make no effort to defend Europe, and forces on the Continent were to withdraw as quickly as possible.[62]

In Asia the Americans would hold the Bering Sea-Japan-Yellow Sea line. Occupation forces in Korea were to retreat to Japan, and forces in China were to move to defensible positions or more probably evacuate the mainland. Four infantry and one airborne division were to defend Japan. Naval forces would keep open communications and blockade Russian ports, and six fighter and two medium bomber groups would protect Japan and participate in the air offensive.[63]

The JSPG then discussed the acquisition of a base area from which to launch air attacks against targets in southern and central Russia. They examined two possibilities--the Cairo-Suez area and northwest India.

In contrast to the November 1947 version, the JSPG expressed grave doubts concerning the viability of the Allied position in the Middle East. To hold Cairo-Suez with the forces available in fiscal 1949 required the planners to accept a number of assumptions and their inherent risks. The planners had to assume that the Soviets would not invade Spain, and not attack Sicily before the Allies occupied the island. They also had to assume that on D-Day the British would have three divisions in Egypt and Palestine, that French North African authorities would cooperate with the United States and that atomic bombs would be used in a tactical role to stem the Red Army's drive into the Balkans and Middle East.[64]

If the assumptions were accepted, the planners outlined minimum force requirements. British forces would defend Egypt and Libya against airborne attacks while the United States sent one infantry and one Marine division to Sicily and an infantry division to North Africa. The Air Force would establish bases along the North African coast from Casablanca to Cairo to assist in the defense of the Mediterranean line of communication. Eleven fighter groups would operate in a defensive role. The Navy would deploy nine carrier task forces to the Mediterranean by D+2 months.[65]

The Allies would next mount a major effort to assist the Turks. Carrier air would supply close air support, and by D+9 months there would be sixteen carrier task forces in the Mediterranean. On the ground three infantry and an armored division would assist the Turks to hold Iskenderon, while five light bomber and nine fighter groups supplied tactical air support. The Air Force would use atomic weapons and strategic air units to assist the tactical forces by striking key bases from which the Soviets staged and supported attacks. The strategic forces would also fly conventional missions against Soviet lines of supply.[66]

The planners then pointed out that requirements and actual forces did not match. There was a serious shortage of anti-aircraft battalions to defend North African bases. The armored division required to help the Turks at Iskenderon was not in fact available. Air Force requirements for defensive missions and for tactical support to the Turks could not be met. For example, of the nine fighter groups needed to support the Turks there was only one available for fiscal 1949. Finally, the tactical use of atomic weapons would seriously reduce the effectiveness of the strategic air offensive.[67]

The planners generally concluded that Cairo-Suez was the best strategic base area in the Middle East. From Cairo-Suez the Air Force could attack targets in southern Russia, and the area was also useful for staging attacks to recover oil resources. Furthermore, Cairo-Suez was well located for amphibious assaults into the Balkans and ultimately into the USSR. The JSPG, however, also realized that given the forces in being, it would be a highly risky strategic venture to try to hold Cairo-Suez and its strategic approaches.

The JSPG, therefore, sought an alternative base area from which to attack target systems on the USSR's southern flank. The Karachi area of India which included Karachi, Lahore, and Peshawar seemed to fit American requirements. From these bases the Air Force could strike southern Russia while the mountainous terrain to the north would make it feasible for relatively small forces to cope with enemy ground offensives. One division, a carrier task force, ten light carriers, and three fighter groups could protect the area by D+2 months. Only small air reinforcements of fighter aircraft would be required in subsequent months.[68]

To protect the line of communications the Americans would need a base in the Casablanca-Port Lyautey area of French Morocco. By D+6 the defense of this region would require one and a third divisions and two and a third air groups.[69]

The JSPG noted that the Karachi alternative had the serious disadvantage of failing to sustain the current policy of supporting countries in the Middle East and Mediterranean. On the other hand, given the limited resources available and the unwillingness to divert atomic and strategic assets to tactical roles, the Karachi alternative appeared to be the only feasible one.[70]

The key to victory was the atomic bomb offensive plus an extensive conventional bombing campaign. The target list remains classified but given the number of weapons available and the fact that the list of industrial target systems to be attacked is almost identical to the target system list in JWPC 486/7, it is reasonable to assume that the twenty-four target cities of 486/7 were also the primary targets of Broiler. Although initial targets were within the USSR, later in the war the Air Force might launch atomic attacks on targets in both satellite and occupied states. The planners noted that industrial targets as well as governmental and scientific complexes could be attacked and destroyed by the simple expedient of attacks on urban areas. The planners also noted that attacks on urban areas would also inflict great psychological damage on the Soviet government and population and perhaps render them incapable of continuing the war.[71]

Execution of the atomic offensive required six groups of B-29s or B-50s by D+1 and twenty groups by D+3. By the latter date the planners noted the need for eight reconnaissance, eight escort fighter, and eleven and a third defense fighter groups. The planners, because of current roles and mission controversies, were unable to decide whether or not naval air should participate in the strategic offensive. A split decision remained in the plan. One view stated that the plan constituted a direct requirement for carrier air and the other that it did not. Finally, the planners noted that chemical and biological warfare could produce useful effects upon the enemy's will to resist. CB warfare would also do minimum damage to economic assets, and the JWPC suggested that the services undertake further study and development of CB warfare.[72]

The planners asserted that the impact of the strategic atomic bombardment campaign and the supplementary conventional air operations might well be the rapid capitulation of the USSR. In case of a quick surrender twenty divisions and twenty fighter groups would occupy twenty-one key Soviet cities. An additional eighteen divisions and eighteen fighter groups would occupy liberated areas and satellite states.[73]

In a sudden shift from the optimistic assessments of the impact of atomic weapons, the JWPC went on to assert that forces available were insufficient to permit full implementation of the strategic air campaign. For D+1 there were five instead of six bomber groups, and by D+3 there would be nine instead of twenty bomber groups.[74] The planners did not go on to describe deficiencies in the number of atomic weapons, aircraft modified to carry A-bombs and trained crews. Nor did the JWPC mention the absence of accurate target data and insufficient training of crews to reach and bomb accurately distant targets at night. They confined themselves to recommending the reckless and dangerous expedient of undertaking the atomic offensive, whatever its problems, as soon as possible after D-Day since conventional alternatives were even less viable.

Although the JWPC was unable to offer an appraisal of the psychological impact of an atomic offensive on Soviet morale they recognized that atomic weapons might not in fact force the Russians to capitulate. Consequently, additional operations would be required after D+9 months.

If the Allies held the Cairo-Suez area, their first major operation would be to regain control of the Middle East's oil resources. U.S. and British forces would advance on Mosul and Baghdad. Thirteen American and five British divisions would be required for the offensive. The Allies would then seize Bahrain and then seek to recover Kuwait.[75]

Having recovered the oil-producing areas, the Allies would next retake Crete and clear the Soviets out of Anatolia. If the Soviets continued to resist, the Allies would liberate Greece, Macedonia and European Turkey as preparatory moves for an invasion of the USSR. By D+18, as air strikes continued against Soviet industry, oil and transportation systems, the Allies would strike into the Black Sea and invade the Ukraine from the south.[76]

If the United States decided to use Northern India rather than Cairo-Suez as one of its main strategic bases, it would still be necessary to recover Middle East oil. The main operations would take place in the Persian Gulf, starting with an amphibious assault on Bandar Abbas followed by the seizure of Bahrain, Kuwait, and Basra. With the Gulf secured, the Allies would then advance up the Tigris-Euphrates valley to Baghdad and Mosul. A second offensive would seek to capture Tehran and reach the line of the Zagros Mountains. Eight British and twelve American divisions supported by eight carrier task forces would participate in the dual offensive.[77]

If the Russians continued to resist, the Allies would proceed to strike the Soviet heartland. The planners presented two possibilities: a thrust via the Mediterranean, Aegean, and Black Sea into the Ukraine, or a drive through the Red Sea and thence north through the Aegean and Black Sea into the Ukraine.[78] Thus, as in the Pincher concept, the Allies would not attempt an invasion of Western Europe, and if an invasion of Russia was required, the main thrust would be directed through the Black Sea into the southern regions of the Soviet Union.

On March 19, 1948, the Joint Logistics Plans Group examined Broiler and noted a number of problems that had to be resolved if the plan was to be feasible. The services had to augment the number of engineer and construction battalions and establish a system to recall

Second World War veterans. An immediate program was necessary to create war reserve stocks and to procure supplies for Allied powers. The services also had to have in place, well before D-Day, operational intermediate bases and loaded ships to supply operations in northern India. The planning group noted that Broiler was mounted on a logistical shoestring. Though basically workable on an austere scale of support, the logistics experts expressed particular worry over strained shipping capabilities.[79] Given the administration's continuing desire to reduce military spending, many of the requirements set forth by the JLPG were probably unrealistic.

In designing Broiler, the planners realized that the United States had limited military capabilities and had to rely heavily on atomic weapons. Atomic strategy had serious flaws. Bombs were in short supply as were modified long-range aircraft and trained crews. Much vital target information was still simply unavailable, and the planners were in fact not able to estimate accurately the impact of an atomic offensive on the Soviet leaders and people. The American military, nevertheless, felt they had no choice but to rely almost exclusively on atomic weapons because they had nothing else.

On March 9, 1948, the Joint Strategic Plans Committee submitted on abbreviated version of Broiler to the JCS. The Chiefs approved it for planning purposes, and a week later the JSPC submitted a revised version designated Frolic for submission to the Secretary of Defense.

The March 9 plan, like Broiler, placed primary emphasis on the use of atomic weapons. The choice of bases included Okinawa, the United Kingdom, and Karachi. The JSPC thus abandoned all hope of holding southern Turkey, Palestine, and Egypt. The planners also designated Iceland as an alternative to bases in England, indicating doubts about Allied ability to protect what had hitherto been regarded as a vital linchpin of Allied strategy.[80]

Plan Frolic of March 17, 1948, included a memorandum for the Secretary of Defense in which the JCS noted that the emergency war plan emphasized the grave military weakness of the United States.[81]

Like Broiler, the plan emphasized the early initiation of a sustained air offensive against the USSR and the recovery of Middle East oil resource areas by the early part of the war's second year. Allied strategy called for the defense of the Western Hemisphere and the United Kingdom, the evacuation of occupation forces from Europe and Korea, securing the Bering Sea-Japan-Yellow Sea line and the initiation of the air offensive from England and Okinawa by D+15 days and from Karachi by D+30 days. The plan assumed that permission to use atomic bombs would be given and that bombs in stockpile would be released to the military.[82]

As in the March 9 plan the Middle East was abandoned as a strategic base area, and Iceland was designated as an alternative to Great Britain. Frolic called for sending troops to Iceland on D-Day or as soon afterward as possible. Forces were also to occupy the Azores and Casablanca. Marines were to be airlifted to Bahrain to assist in the evacuation of U.S. nationals and possibly to destroy oil installations, although this would also deny them to the Allies for a long time after their recapture.[83]

The plan noted that American strategy had numerous shortcomings, including the failure to assist the nations of Western Europe and the inability to retain Middle Eastern oil resources.[84] Risks and shortcomings, however, had to be accepted because of the absence of sufficient forces. On May 4, 1948, the JSPG drew-up directives for the implementation of Frolic, and on June 25 the Joint Logistics Plans Committee submitted a directive for the development of logistics plans in support of Frolic.[85]

The individual service chiefs were not, however, fully satisfied with Frolic. On April 6, 1948, the Chief of Naval Operations, Admiral Denfeld, sent a critical memorandum to the other Chiefs. He noted that the plan gave no encouragement to the European democracies and that Frolic should not be used for anything but short-range planning. It was, he stated, inadequate for medium-range planning.[86]

Among other deficiencies in the plan, Admiral Denfeld noted that in addition to abandoning Europe without a struggle Frolic accepted the loss of the Mediterranean Sea, offered virtually no assistance to the Turks and did not safeguard Middle East oil. Moreover, by relying entirely upon atomic strikes for the defeat of the USSR the Allies made no provision for action in the event of failure or only partial success of the air-atomic strategy. If the atomic offensive did not succeed, the Allies would have in the meantime lost so much territory and so many resources that the prospect of ultimate victory would be jeopardized. Finally, the CNO asserted that Karachi was not a useful base. To support forces at Karachi would require a tremendous logistical effort, and a base at Karachi could not give support to or be supported by any other operations in the plan.[87]

Admiral Denfeld also asserted that the Soviets were not as strong nor the Americans as weak as Frolic presumed. The threat of subversion was, he believed, drastically overstated, and the plan as a whole was based on America's worst fears, present shortcomings and glaring weaknesses. He also raised doubts as to whether the Russians would in fact capitulate after an atomic offensive and wondered if the U.S. military would indeed receive permission to employ A-bombs at the start of a war.[88]

The Admiral also submitted a different concept for mid-range planning. He advocated holding the Russians along the line of the Rhine and securing Allied positions in Cairo-Suez and Basra. Allied defense would be primarily by conventional means, and atomic weapons would be used selectively against tactical targets that would help to halt the Red Army.[89]

On April 27, General Spaatz, Chief of Staff of the Air Force, responded to Admiral Denfeld's paper. Spaatz felt that Frolic should be approved as an immediate emergency plan based on current capabilities. Spaatz did, however, agree with the CNO that additional planning based on enhanced strength and on the willingness of Western European nations to resist the USSR should be set in motion.[90]

The CNO, on August 11, 1948, offered additional criticisms of Frolic. The plan, he noted, called for operations on widely separated fronts which were not mutually supporting. Each operational area required a separate line of communication which would have to be protected, thereby requiring the use of a disproportionate amount of America's total military resources. The establishment of a heavy bomber air base system in the Karachi area required such a heavy drain on air and sea transport assets that operations in all areas would be seriously reduced. Sustaining forces in north India would place a continuing drain on these resources. Finally, the CNO noted that aside from the bombing offensive further operations from Karachi were not feasible, and if offensive operations were required, the Allies would have to reopen the Mediterranean which would be firmly held and strongly defended by the Soviets. Sustaining Karachi and opening the Mediterranean would probably not be feasible.[91]

On September 3, 1948, General Bradley, the Army Chief of Staff, defended Frolic. He pointed out that the Soviet air and sea threat to a LOC from the Cape of Good Hope to Karachi would be less serious than to a Mediterranean LOC. General Bradley further noted that

Britain's position in the Middle East was deteriorating, thereby increasing the risks to any effort to hold the Mediterranean. The United States, therefore, required a plan that did not necessitate immediate access to the Mediterranean or the Middle East. Bradley admitted that Frolic was not strategically sound but concluded that since the Air Force required a base complex in the Karachi-Lahore area, planning for such a base should continue on a low-priority basis.[92] Admiral Denfeld responded on September 22 that planning for a Karachi-Lahore base for use as a last resort was a reasonable precaution, but he called again for a plan that provided for retaining control of the Mediterranean.[93]

None of the Chiefs were fully satisfied with Frolic. At best they regarded it as marginally acceptable as an immediate emergency plan that reflected America's lack of readily available military resources. They, therefore, agreed to revise the plan based on hopes for a larger defense budget and upon conversations with Canada and Great Britain. Increased Western European strength was another factor that the JCS hoped would make a difference in their strategic planning.

The American military establishment during 1947-1948 had to work without detailed political guidance. Moreover, they regarded their forces in being as inadequate to cope with Soviet conventional forces in a major war which they assumed would be global and require a complete military and political victory. In the absence of sufficient conventional forces the military felt they had to rely almost completely on atomic weapons to turn the tide of battle despite the severe limitations that American strategic force levels placed upon atomic warfare. The planners had to assume with no assurance that there were sufficient bombs, planes and trained crews to carry out the A-bomb offensive. They also had to assume that they would have political approval to use atomic bombs in the opening days of any war.

In contrast to the Pincher plans and the various mobilization studies, the Broiler and Frolic concepts not only abandoned Western Europe, North China, and Korea to the Soviets but also gave up efforts to hold southern Turkey, the Syrian coast, Palestine, and Cairo-Suez. In fact, the JSPG relinquished the entire Mediterranean basin without serious resistance. The planners even implied doubts about the ability of Anglo-American forces to secure the British Isles as a strategic base complex.

The concept that a war with the USSR would be total and global and that American use of the atomic bomb would counter Soviet conventional superiority had by early 1948 come to dominate American strategic thinking. The planners did not think that Moscow would launch a limited war using Russian or surrogate forces. Any war would be total and fought for world domination. The view of total global atomic war was to dominate both short- and long-range planning for many years.

NOTES

1. James F. Schnabel. *The History of the Joint Chiefs of Staff. The Joint Chiefs of Staff and National Policy, Volume I, 1945–1947.* Wilmington, Delaware: Michael Glazier, Inc., 1979, pp.164–165.
2. JIS 267/1, 14 January 1947.
3. JIC 237, 17 January 1947.
4. JIS 275/1, 18 April 1947.
5. JIC 374/2, Capabilities and Military Potential of Soviet and non-Soviet Powers in 1956. 8 January 1947.
6. JCS 1770, 28 April 1947.
7. JCS 1770/1, 7 October 1947.
8. JCS 1725/1, Strategic Guidance for Industrial Mobilization Planning. 13 February 1947.
9. *Ibid.*, Appendix A.
10. *Ibid.*
11. *Ibid.*
12. *Ibid.*
13. *Ibid.*
14. *Ibid.*, Annex A to Appendix B.
15. *Ibid.*, Annex B to Appendix B.
16. *Ibid.*, Annex C to Appendix B.
17. JWPC 486/7, 29 July 1947.
18. *Ibid.*
19. *Ibid.*
20. *Ibid.*
21. *Ibid.*
22. *Ibid.*
23. *Ibid.*
24. *Ibid.*, Tab A, Tab B, and Tab C.

24. *Ibid.*, Tab A, Tab B, and Tab C.
25. JWPC 486/8, Guidance for Mobilization Planning as affected by the loss of the Mediterranean Line of Communications. 18 August 1947.
26. *Ibid.*
27. *Ibid.*
28. *Ibid.*
29. *Ibid.*
30. JLPC 17/11, 1 September 1947.
31. JCS 1796/1, Aircraft Requirements for War in the Proximate Future. 14 October 1947.
32. JCS 1725/3, 14 November 1947.
33. JLCP 416/8, 3 December 1947.
34. JLPC 416/9, 3 December 1947.
35. JSPG 499/2, Charioteer, 3 December 1947. The JSPG replaced the JPS under the National Security Act.
36. *Ibid.*
37. *Ibid.*
38. *Ibid.*
39. *Ibid.*
40. *Ibid.*
41. *Ibid.*
42. JSPC 846/7, 3 December 1947. *See also*: JCS 1796/6, 10 December 1947.
43. JSPG 496/1, Broiler, 8 November 1947.
44. *Ibid.*
45. JSPG 496/4, Broiler, 11 February 1948.
46. *Ibid.*, Appendix Part IV.
47. *Ibid.*, Appendix Annex A, Part IV.
48. *Ibid.*
49. *Ibid.*
50. *Ibid.*, Appendix Annex A, Part V. See also Tab A.
51. *Ibid.*, Tab B.
52. *Ibid.*, Annex A, Part VI.
53. *Ibid.*
54. *Ibid.*
55. *Ibid.*
56. *Ibid.*
57. *Ibid.*

58. *Ibid.*, Annex A, Tab D, Table A.
59. *Ibid.*
60. *Ibid.*
61. *Ibid.*, Appendix Annex B.
62. *Ibid.*
63. *Ibid.*
64. *Ibid.*, Appendix Annex A, Part X.
65. *Ibid.*
66. *Ibid.*
67. *Ibid.*, Appendix Annex B.
68. *Ibid.*
69. *Ibid.*
70. *Ibid.*, Appendix Annex A, Part X.
71. *Ibid.*, Appendix Annex C.
72. *Ibid.*
73. *Ibid.*, Appendix Annex B.
74. *Ibid.*, Appendix Annex B, Table B–1.
75. *Ibid.*, Appendix Annex A, Part X.
76. *Ibid.*
77. *Ibid.*, Appendix Annex A and Annex B.
78. *Ibid.*
79. JLPG 84/5, Quick Feasibility Test of JSPG 496/4, 19 March 1948.
80. JCS 1844, 9 March 1948.
81. JCS 1844/1, Short Range Emergency Plan. Short Title: Frolic, 17 March 1948. On April 5, 1949, *Frolic*'s title was changed to *Grabber*.
82. *Ibid.*
83. *Ibid.*
84. *Ibid.*
85. JSPG 496/11, 4 May 1948 and JLPC 416/4, 25 June 1948.
86. JCS 1844/2, 6 April 1948.
87. *Ibid.*
88. *Ibid.*
89. *Ibid.*
90. JCS 1844/3, 27 April 1948.
91. JCS 1844/14, 11 August 1948.
92. JCS 1844/20, 3 September 1948.
93. JCS 1844/26, 22 September 1948.

CHAPTER IV

From Bushwacker *to* Halfmoon

By the end of 1947 the United States had become deeply engaged in European and global power politics. Government leaders had come to the conclusions that the Soviet Union was a ruthless enemy pursuing a policy of relentless expansion and only the United States could check the USSR's drive for domination. President Truman and his advisors doubtless thought that the main Soviet threats to the rest of the world's nations were political and economic, but such views were cold comfort to the military professionals who had to devise plans to cope with the unlikely contingency that the Cold War would suddenly become a real conflict.

The Truman administration continued to place severe restraints upon defense spending. Retrenchment rather than rearmament was a permanent feature of government defense policies. The Joint Chiefs of Staff thus faced a continuing predicament—devising military strategies that required the armed services to meet expanding commitments with static and insufficient forces. Given current and projected defense budgets, military leaders could foresee no clear or rapid solution to their dilemma.

Early in 1948 a detailed examination of Plan Charioteer clearly indicated that the gap between assets and requirements was serious. Moreover, the analysis of the plan seemed to prove that the immediate future held little promise of improvement. At current levels of defense spending, the situation in 1955–56 would in all probability be no better than it was in 1947–48.

In addition to describing the opening phases of a general war, Plan Charioteer was especially concerned with aircraft requirements during the first six months of hostilities.[1] Since the strategy of Charioteer closely resembled the concepts of Broiler and Frolic, the planners assumed that vastly superior Soviet ground and air forces would mount several offensive operations and quickly overrun most of Western Europe and the Middle East.[2] The United States would have to rely primarily on strategic and tactical air power.

At the start of hostilities the Air Force would begin a strategic air offensive from American bases with long-range B-36 bombers. B-29s and B-50s would join the attack as soon as the armed forces established and secured forward bases.[3] Tactical air units would protect vital targets in the United States, escort bombers for part of their missions, assist in defending forward bases, and help guard lines of communications.[4] By D+6 months the Navy would have to deploy 3,979 aircraft and maintain 5,211 replacements.[5] The frontline strength of the Air Force by D+6 months would grow to 166-1/3 groups and twenty-eight independent squadrons with a total of 7,538 aircraft.[6] Ground and sea forces also required expansion in order to complete their missions. Given an administration determined to keep down defense expenditures, the force projections for both D-Day and D+6 months were probably overly optimistic.

A Munitions Board report of January 23, 1948, indicated that Broiler, Frolic, and Charioteer called for resources that were simply not available.[7] The Board, in fact, recommended that the JCS produce a new, more realistic plan based on a cautious estimate of force levels and capabilities.[8] On February 18, the Munitions Board noted that for fiscal 1949 the military required $139.6 million for stockpiling and for creating stand-by plants, but that the budget allocation was only $37.6 million.[9]

Despite the substantial gap between desired and available forces the planners submitted another long-range war plan in early March. Plan Bushwacker of March 8, 1948, assumed that war was forced upon the United States and its allies, Great Britain and the Commonwealth, on or about January 1, 1952.[10]

The planners assumed that by 1952 the USSR would have greatly enhanced military power. The army would consist of 110 divisions in being.[11] After 180 days of mobilization the Red Army could place 500 divisions in the field backed by 13,000 self-propelled guns, 77,500 towed artillery pieces, 7,500 rocket launchers, and over 30,000 tanks. Most of the equipment would consist of post-war production models.[12] The navy would contain 200 high-speed modern submarines and 130 older, modernized U-boats.[13] The Red Air Force would possess 20,000 combat aircraft largely jet propelled and a long-range bomber force of about 1,600 planes, roughly equivalent in range and bomb load to the B-29.[14]

The JSPG assumed that the Soviet Union would not have atomic bombs in 1952 but would possess other atomic munitions in limited quantity. The planners did not describe these munitions in any detail, mentioning only that these weapons would probably be radiological and air delivered.[15] There was no discussion of the nature of these weapons or of their destructive capabilities. All the planners noted was that these other atomic munitions in conjunction with biological warfare posed a significant threat to the American industrial base.

The planners did not provide a projection of the political situation in 1952. They evidently presumed that there would be no fundamental changes. Thus Western Europe, despite American aid, would remain weak and divided. The Civil War in China would still be unresolved, and Great Britain would maintain its strong imperial position.

As with earlier plans Bushwacker presumed that the USSR did not seek war with the United States, at least for the moment, and that if hostilities took place, the cause would be Soviet miscalculation of Washington's willingness to resist piecemeal aggression. Once hostilities began, the JSPG assumed that the Red Army would overrun wide areas in Western Europe, the Middle East, and Far East. Bushwacker did not provide any description of probable Soviet campaigns on the Eurasian mainland, but the plan did note that Russian mobilization would probably provide the United States with strategic warning.[16]

Enhanced Soviet military capabilities would allow the Russians to exercise strategic options that earlier plans had assumed were not feasible. The JSPG assumed that the British could defend the United Kingdom, but the Joint Intelligence Group disagreed. The JIG believed that the USSR could render the British Isles impractical for use as an advanced base for strategic air attacks against the Soviet homeland. The two groups were unable to resolve their dispute, and a split decision was recorded in the plan. The JSPG and JIG did, however, agree that the Red Air Force could attack strategic targets in Japan and Alaska and that unless the United States reacted swiftly, the Russians, using airborne forces, would be able to seize or neutralize Iceland and Greenland. The Soviets would attack the continental United States with their mysterious atomic munitions and other undefined unconventional weapons, and Moscow would also mount a major sabotage and subversion effort.[17]

The Western Allies would seek as their basic war aim to create conditions within the USSR that would assure the abandonment of political and military aggression. Bushwacker did not call for unconditional surrender, but given the stated political goal, nothing short of the overthrow of the Soviet regime would have produced the desired results.[18]

American strategy called for the protection of essential war industries within the United States, the securing and defense of strategic bases and their lines of communications and the launching of a strategic atomic offensive against the USSR. The United States would also attempt to provide military assistance to Allied forces, destroy enemy air and naval forces, undertake psychological warfare operations against the Soviets, and mobilize additional forces in case the atomic offensive failed to force the Soviets to capitulate.[19]

The planners presumed that the war would unfold in phases. Phase I would last for sixteen days during which time the American armed forces would devote themselves to protecting vital targets from Russian attack. The second phase involved the Allied offensives, and the final phase extended from the peak of the Allied offensives to the end of hostilities. The planners did not estimate how long the second and third phases would last, nor did they specify what ground, sea and air operations the Allies would launch.[20] To predict the course and timing of a war is difficult if not impossible, and the JSPG confined itself to noting that the Allies would have to reoccupy or retain Persian Gulf oil-producing regions and create a vast forward base structure for the air assault and other offensive operations.

The plan also emphasized the critical importance of having forces ready for instant reaction to Soviet moves. Virtually all of the military operations during the first two phases of hostilities had to be executed by those forces currently under arms. Forces mobilized after D-Day would be ready only in the war's final phase.

In defending the productive capacities of the continental United States the planners concluded that it was not possible to protect the thousands of targets that the Russians might choose to attack. Only the most critical facilities could be covered, and the most important single complex comprised the plans and facilities which produced, stored, and assembled atomic bombs. There were a total of seven sites which required three division equivalents, forty-two anti-aircraft battalions and three fighter groups for their defense. The Army required for the first phases of the war a total of fourteen and one-third divisions and over three hundred anti-aircraft battalions ready for immediate action on D-Day.[21] The Navy required eight carriers, and the Air Force needed seventeen fighter and eight heavy bomber groups.[22]

On D-Day or even earlier, provided that sufficient warning were available, a fast-carrier task force would proceed to the Greenland-Iceland area to check any Soviet airborne assaults against either island. A regimental combat team would proceed by air to Iceland followed by two more regiments sent by sea. An additional regiment would proceed by sea to Greenland. Fifteen anti-aircraft battalions and four fighter groups would reinforce the original garrison. The Americans would occupy Iceland and Greenland whether or not they had received the permissions of the Icelandic government and Danish authorities. Once secure, Iceland and Greenland would become strategic air bases, and forces deployed there would join the atomic aerial offensive which had already begun from bases in the continental United States, Alaska, and Okinawa.[23] Additional bases were to be developed in the Gold Coast of Africa, Sudan, Aden, and Northern India after D-Day. If the United Kingdom was able to fend-off Soviet air attacks, the Air Force would also deploy strategic bombers to the British Isles.[24]

The basic objectives of the air offensive were to reduce immediately Soviet offensive capabilities and simultaneously reduce the USSR's will to resist by inflicting heavy casualties upon the civilian population and causing major disruptions of the social and economic system. Immediate and crippling attacks, launched by ready forces from existing facilities, would begin within hours after D-Day. Despite a continuing lack of precise target data for many regions of the USSR, the planners concluded that the most important targets were: stockpiles of atomic munitions and the facilities for their production, key government and control facilities, urban industrial areas, the oil industry, submarine bases, construction and repair facilities, the transportation system, the aircraft industry, the coke, iron and steel industry, and the electrical power system.[25]

Large percentages of these complexes could be struck by attacking twenty major urban areas, including: Moscow, Leningrad, Gorky, Kuybyshev, Baku, Ufa, Sverdlovsk, Niszhniy Tagil, Novosibirsk, Stalinsk, Molotov, Saratov, Kazan, Chelyabinsk, Omsk, Kharkov, Yaroslave, Grosny, Stalingrad and Magnitogorsk. These cities contained 17 million people, about ten percent of the population, and produced ninety-nine percent of Russia's combat aircraft engines, eighty-six percent of the airframes, ninety-nine percent of the tanks and SP guns, and sixty-one percent of the USSR's refining capacity.[26]

Tactically, bombers carrying atomic bombs would operate singly or in pairs accompanied by four to eight reconnaissance aircraft and several additional bombers for diversion and saturation of defenses. Air refueling would be a practical technique by 1952 and would allow B-50 and B-36 aircraft to reach all of the designated targets in the USSR. The atomic campaign would be supplemented by conventional bombing operations and biological warfare strikes. As the air force developed its forward bases the intensity and tempo of the assaults would increase. Ultimately, seventeen heavy and eleven medium bomber groups would strike the USSR from a global network of bases.[27]

Continuing interservice rivalries over roles and missions prevented the planners from defining the navy's role in the strategic air campaign. The Navy asserted that carrier aircraft could participate in the atomic offensive and that carrier planes could deliver atomic weapons. Naval air could also conduct conventional bombing operations. A single air coordinator should, therefore, plan for joint Air Force-Navy strategic air operations. The Air Force argued that the strategic bombing mission should be a single-service operation. In a war the Navy would have enough to do without participation in the atomic offensive. Later, however, carrier forces could participate in conventional bombing operations, especially against naval and coastal targets. Unable to resolve the conflicting views, the planners had no choice but to record a split decision.[28]

The United States would also seek to hold or regain as soon as possible Persian Gulf oil-resource regions. Action in the Gulf would be closely coordinated with the British. The Allies would delay Soviet moves toward the Gulf by sabotaging road and rail links between the USSR and the Gulf. Sabotage operations would be planned in advance and executed starting on D-Day. The Navy would supplement the sabotage effort by launching carrier strikes from permanently stationed task forces in the eastern Mediterranean. The U.S. and Britain would also maintain a minimum of two regimental combat teams and a fighter group in the Middle East. This force would be available for deployment to the Gulf region prior to D-Day, if possible, or else on D-Day. Ready reinforcements, consisting of an infantry, an armored and an air-mobile division and three air groups, would begin moving to the Gulf on or if possible before D-Day. By D+6 months three British and ten American divisions would be operating in the Gulf region.[29]

If the aerial offensive failed to secure a Soviet capitulation, the Allies would have to undertake further operations. The first step would be to seize or occupy advanced bases in Turkey, Scandinavia, Great Britain, and North Africa. From these bases the air offensive would be able to strike at all important Soviet economic and industrial complexes. The Allies would also seize advanced bases for the purpose of mounting an invasion of the USSR, but the staff planners did not describe the subsequent campaigns. Nevertheless, to achieve these objectives the Army would have to expand to twenty-nine divisions by D+18 months and ultimately reach its World War II peak of ninety divisions. The Navy by D+18 would have to deploy twenty-four carriers and the Air Force would expand to seventy-five groups.[30]

Although the planners rejected the idea of forcing unconditional surrender upon the USSR, they drew up plans to occupy the Soviet state and its satellites, allotting forty-four air groups and twenty-five divisions for the task.[31] The JSPG did not explain the seeming contradiction in their plan, nor did they explain how the Allies could in fact occupy cities reduced to chaos and rubble under atomic and conventional bombing attacks.

In writing Bushwacker, the planners followed earlier operational concepts by assuming that Soviet forces could quickly capture continental Europe, much of the Middle East, parts of China and Korea. The JSPG also presumed that as in the Broiler-Frolic plan Russian forces could overrun Turkey and Cairo-Suez and neutralize the British Isles. Iceland and Greenland replaced the United Kingdom as the preferred forward bases in the Atlantic area. The resort to an atomic counteroffensive as soon as hostilities began was also derived from earlier plans, although the question of presidential release authority remained unresolved. The planners assumed but did not, in fact, know that the president would permit atomic warfare within hours of the start of hostilities. The new element in Bushwacker was the requirement for substantial ready forces capable of going into action on or before D-Day. The size of these forces and the expense of maintaining them in an operational posture would have required a much increased defense budget, an unlikely possibility in 1948 or in the years following. In any event the Chiefs, given their current budget and roles and missions problems, did not attempt to act on Bushwacker.

The Chiefs and their staffs also continued trying to produce an emergency plan acceptable to all the services. On May 11, 1948, the Joint Strategic Plans Group sent to the Joint Strategic Plans Committee its revised version, named Crankshaft, of the Broiler-Frolic emergency war plan. Crankshaft assumed that war was forced upon the United States during fiscal year 1949, and the forces involved were those actually existing in 1948–1949.[32] The planners explicitly noted that the state of American and British forces was by no means ideal and that the plan should definitely not serve as a basis for budgetary or long-range mobilization planning.[33]

As in earlier plans the JSPG assumed that the USSR enjoyed a great numerical advantage in ground and tactical air forces. In July 1948 the planners credited the USSR with an army of 135 divisions supported by 100 satellite divisions, including Finnish forces. The Red Air Force consisted of 13,000 combat planes plus about 1,400 satellite aircraft. The main naval threat consisted of over 200 submarines. The U.S. fielded nine divisions and nine separate regiments, most of which were not ready for combat, eleven carriers, and fifty-three and two-thirds air groups plus thirteen separate squadrons. Britain and her empire and dominions possessed twelve divisions and thirty-one brigades, eight carriers and 1,900 combat aircraft.[34]

Given the imbalance of forces, the JSPG believed that the western Allies could not halt the initial Soviet onslaught and to try to do so would lead to defeat in detail and the destruction of the relatively small combat-ready Allied forces. Therefore, the planners intended to concede Western Europe, the Middle East, the Persian Gulf, north and central China, and South Korea. The defense of these regions was to be left to local forces. The Allies would provide some air support and assist in executing a demolition plan to deny the Soviets industrial resources and access to transportation systems.[35]

The Russians would be able to overrun Western Europe to the Pyrenees by D+60 days. The planners disagreed over whether or not the Soviets would then invade Spain, and recorded a split decision on this issue. Yugoslav forces, backed later by Russian divisions, would seize mainland Italy by D+75 days, and Sicily would fall by D+105 days. Satellite forces would overrun Greece by D+60 to D+90 days, depending upon the effectiveness of demolitions of the road and rail net. In forty-five additional days the Aegean Islands and Crete would fall to the Soviet and Satellite forces. Turkey would hold out somewhat longer, but the Red Army would be in Aleppo by D+150. By D+175 days the JSPG expected the Red Army to capture Suez. In the Gulf region the Soviets by D+30 days would have gained control of the Basra-Abadan area and by D+60 would control all of Iran and Iraq. In the Far East the Soviets would take China to the Yellow River and South Korea by D+150. The Russians would also close the Mediterranean line of communication from air bases established in Sicily and would launch a major air assault on England by D+60. The JSPG believed that England could effectively defend itself. The Joint Intelligence Committee disagreed, and the difference of opinion remained unresolved in the plan.[36]

The American-Allied response called for the establishment and defense of strategic base areas and the launching of an atomic bomb offensive. Designated base areas included the continental United States, the United Kingdom, Okinawa, and Karachi. Bases in the Cairo-Suez and Basra areas were viewed as being indefensible. Cognizant of previous criticisms by the Navy, the JSPG admitted that Karachi was far from being an ideal base. The area was, however, easily defensible and thus useful in an emergency when forces were severely limited. The targets for the atomic offensive were essentially urban areas. Their destruction, the planners believed, would seriously disrupt the USSR's war-making potential and undermine Russian morale. Conventional bombing would supplement atomic attacks, and if resistance continued, atomic and conventional strikes would be mounted against targets in satellite states and in occupied areas. The Allies would also evacuate troops from Europe and take measures to guard vital industrial complexes from subversives, who, according to the plan, constituted a major threat to Western security.[37]

After providing for the defense of vital economic-industrial resources, securing strategic bases and mounting the air offensive, a process that would take about nine months, the JSPG felt that the USSR might capitulate under the hammer blows of the strategic bomber force. If, however, the atomic offensive was not as effective as predicted, or if the USSR decided to fight on despite the tremendous toll caused by the rain of atomic weapons upon the Soviet homeland, the Allies would have to undertake conventional offensive operations.

The first call upon Allied resources would be the recapture of Persian Gulf oil resources. The planners assumed that the Western powers required Middle East oil to fight a prolonged conflict. In the first phase of the conflict, however, the JSPG called for the dispatch of a Marine battalion to Bahrein. The Marines were to destroy the oil facilities while the British neutralized the oil fields in the Basra and Kirkuk areas.[38] If driven out of the region, the Russians would probably carry out additional demolitions, creating a level of destruction that would render the oil fields inoperable for an extended period. Thus, a successful Allied counteroffensive might not in fact place Anglo-American forces in a position to use Gulf oil or at very least require a massive and lengthy repair effort. The Western powers had not yet fully exploited the full potential of Saudi oil fields, but this required a major long-term effort that would not bear fruit for many years. A successful counteroffensive in the Gulf seemed to promise only the recapture of unuseable resources, but despite this seeming contradiction the JSPG advocated a major effort in the Gulf region.

The Allies would concentrate forces on the west coast of India and possibly in Aden and East Africa. Carrier air and land-based air from Karachi would support an amphibious assault on Bandar Abbas. After further reinforcement of the Bandar Abbas area, the Allies would seize Qatif and Bahrein. Forces would again move forward and launch a major blow at the Kuwait-Basra area, thereby securing a beachhead at the head of the Persian Gulf. Allied columns would then advance up the Tigris-Euphrates Valley and along the railway and road lines to Tehran, driving the Soviets back to the mountains in Turkey and northern Iran.[39] The planners evidently assumed that logistical support for two major offensives moving in divergent directions was somehow feasible.

For the offensives the British would contribute nine divisions, fifteen fighter squadrons, and two light carriers. The Americans would send eighteen divisions, eight carrier task groups, and fifteen fighter and seven bomber groups.[40]

Additional offensive operations called for the Allies to seize a base area in the Casablanca-Port Lyautey region of French Morocco. Forces operating from this base complex would be projected, according to circumstances, into the Mediterranean, Spain, or northwest Europe. Operations in the Mediterranean would be directed at the reconquest of the Aegean, and from there the Allied forces would advance via the Black Sea into the Ukraine. Another thrust would come through the Red Sea to open the Suez Canal, and from there the Allies would also advance into southern Russia via the Aegean and Black Seas. The JSPG regarded a continuation of the Gulf offensive into Russia as too difficult logistically and tactically but asserted the offensives via western Europe, the Mediterranean and Red Sea, either singly or in combination, were within Allied capabilities.[41]

The strategic problems involved in mounting the Gulf offensive and then halting it and sustaining the forces in place or redeploying them were not addressed. Finally, the planners noted that if all the projected efforts failed to force the USSR to capitulate, a major invasion of the Soviet Union would be necessary. Such an invasion would put such strains on Allied resources that it would be unsupportable. Other methods were, therefore, necessary to force Moscow to surrender, but the planners did not and perhaps could not describe them.[42]

Plan Crankshaft recognized that if the atomic offensive failed to produce quick-decisive results, the Allies would face the bleak prospect of a long war. Conventional strategic options provided no solution to regaining in useable form vital oil resources. Nor did the strategic options take into account the difficulties of sustaining a multi-front war in such a manner that the individual operations could be effectively coordinated in pursuit of a single strategic design.

While the planners worked, the JCS kept the British and Canadians informed of their progress. The mechanisms of coordination established in World War II continued to function on an informal basis, supplemented by periodic "official" conferences. The special relationship between Washington and London was real, and from April 12 to 21 planning officers from England, Canada, and the United States met in the American capitol to discuss Broiler-Frolic-Crankshaft. They also accepted a new plan, Halfmoon, as a basis for the development of separate but coordinated plans by each participant. The JCS approved Halfmoon as guidance for the military departments and for the unified and specified commanders on May 19, 1948.[43]

Halfmoon was a short-range emergency war plan and dealt primarily with the first year of hostilities. Its strategic concepts followed closely those of the Broiler series with modifications introduced as a result of interservice and interallied discussions. The plan responded to the premise that the USSR attacked the United States and its Allies during fiscal year 1949. Moscow's attack would be the result of a miscalculation of the vigor that the West would display in resisting Soviet expansion or a preemptive strike to prevent the Allies from creating an effective defensive shield.[44]

The Chiefs noted that the political authorities had still not offered a definitive set of political goals. They, therefore, presumed that Allied war aims would include at a minimum the reduction of the USSR to its 1939 boundaries and the creation of conditions within the Soviet Union which would assure the abandonment of political and military aggression.[45] As in earlier plans, the United States did not demand unconditional surrender but did establish aims that required complete capitulation for their accomplishment.

The JCS presumed that England, the Dominions and the British colonial empire would fight with the United States. France and the Benelux countries would also side actively with the Allies. Turkey, Spain, and Norway would resist if attacked while Sweden and Denmark would seek to remain neutral and would probably submit if attacked by the Red Army. Middle Eastern countries would ally with the United States but would not be able to offer effective military resistance while Latin America in joining the Allies would be able to maintain its own internal security.[46]

The Soviets at the start of hostilities would launch offensives to capture the Middle East and its oil resources, overrun Western Europe, neutralize the United Kingdom, expand positions in China and Korea, disrupt Allied war-making capacity by subversion and sabotage, and disrupt Allied communications lines by submarine warfare, mining, and air operations.[47]

The Allied plan of action called for mobilization, the early initiation of an air offensive against vital elements of the USSR's war industries and the regaining of Middle East oil resources. Allied offensive efforts would thus focus on western Eurasia. In the Far East the Allies, except for the aerial offensive, would stand on the strategic defensive.[48]

At the start of hostilities the Allies would first seek to secure the Western Hemisphere and its resources by the activation of a limited air-warning net. Canada would provide an infantry brigade group and two light-bomber squadrons for the defense of northeastern Canada. Six American fighter groups and two-and-a-half Canadian squadrons would defend critical industrial areas. An infantry regiment would move north to defend Fairbanks and Anchorage, and two airborne regiments with necessary transport squadrons would act as a strategic reserve. A regiment would guard the Sault Ste Marie Canal, and three division equivalents would guard other essential sites against sabotage. Two regiments and two fighter squadrons would secure the Panama Canal and other Caribbean bases.[49]

Allied occupation forces in Europe would withdraw from the Continent, but in contrast to earlier plans the Allies would conduct a fighting retreat in some areas. In Germany the Allies would retreat to the Rhine, seek to delay further Russian advances and stimulate resistance movements. From the Rhine the British would proceed to Dunkirk and thence to England. American forces would withdraw to France and fall back, fighting delaying actions, either to French coastal ports or the Pyrenees. U.S. forces in Austria were, if possible, to join American units in Germany. If the Russian advance were too swift, the troops would retreat through the Belfort Gap or Italy or in the worst case enter Switzerland. British forces in Austria would retreat through Germany or Italy depending upon circumstances. Allied forces in Trieste and Greece would not seek to offer extended resistance. Allied naval forces would evacuate them.[50] Thus, for the first time, the JCS accepted the idea of fighting in Western Europe. The Americans and British did not believe they could halt the Red Army but had, nevertheless, decided not to abandon the Continent without significant resistance.

The Allies would also secure the United Kingdom. During the first six months the British would have to act alone. The United Kingdom on D-Day would be able to deploy twenty-nine battalions, twenty anti-aircraft regiments, 192 fighters and 160 bombers. By D+6 months the British would have 103 battalions, sixty anti-aircraft regiments and 520 fighters. Two light carriers would also be available by D+1 month.[51] After D+6 months the United States would if necessary supply additional air and anti-aircraft units.

At the beginning of the war the Allies would immediately secure the Cairo-Suez area as a strategic base. Khartoum was included in the Cairo-Suez base region. The planners dropped their call for a major base at Karachi due to pressure from the Navy and the British. On D-Day the British would have a division and an Arab Legion regiment in the area, and by D+6 months three and two-thirds British divisions would protect the region. Ninety-six fighters and twenty-four bombers would be in Cairo-Suez on D-Day. Six months into the war the RAF would deploy 168 fighters and 200 bombers. By D+6 months the U.S. would have reinforced the British with two and two-thirds divisions and 589 fighters. The Cairo-Suez base would receive supplies via the Mediterranean, and if and when the Soviets closed the Mediterranean route the Allies would establish a new line of communications via the Cape of Good Hope and the Red Sea. The JCS, due to lack of sufficient forces, could offer no assistance in the defense of Spain, Sicily, French North Africa, or Malta. Four American carriers and four British light carriers would offer some protection to the Mediterranean line of communications, but the JCS expected to have to rely on the longer Cape-Red Sea route by D+6 months.[52]

In the Far East American forces would seek to protect the Bering Sea-Japan-Yellow Sea line. The United States would withdraw its forces from Korea and China, defend Japan and, if feasible, provide military assistance to Nationalist China.[53]

During the opening months of the war the Air Force would launch an air offensive using atomic weapons against the USSR. The JCS presumed that they would receive political permission to employ atomic bombs and intended to begin operations from bases in England, Cairo-Suez, and Okinawa on D+15 days.[54] The plan provided no detailed target list but given the facts that the atomic stockpile was in the range of fifty weapons and Halfmoon derived from Broiler, it is reasonable to assume that the targets included the twenty cities mentioned in earlier studies. A mobilization plan written to conform to Halfmoon in fact noted that the Strategic Air Command would strike twenty Soviet cities with atomic bombs. In conjunction with the atomic campaign the Strategic Air Command would also mount conventional operations against Russian target complexes.

Additional initial operations included the establishment of air and naval bases in Iceland and the Azores on D-Day or as soon afterward as possible. A Marine battalion would deploy to Bahrein to assist in the evacuation of American nationals and to neutralize oil installations. The British would simultaneously destroy oil facilities in Iraq and at the head of the Persian Gulf and sabotage rail lines leading to the Gulf.[55]

After D+6 months the Allies would strengthen the defense of strategic base areas, improve and expand the base facilities and intensify the air offensive. By D+9 an additional three American and one British division plus six and two-thirds American fighter groups would be deployed in the defense of Cairo-Suez-Khartoum. Reinforcements for the defense of England would depend upon the effectiveness of the Soviet air assault and British air defense. In addition, the Allies would have by D+12 an additional twenty-three divisions and 1,400 aircraft for use in counteroffensives.[56]

Possible operations included the reopening of the Mediterranean and the regaining of Middle East oil. If the air offensive reduced Soviet capabilities and if Spain were unoccupied, the JCS felt that the Allies could regain Sicily, establish strong air forces in North Africa, and reopen Algerian and Tunisian ports. The Allies also had to recapture Middle East oil resources by D+24 months. Allied offensives might be directed from Cairo-Suez towards Kirkuk, against the Persian Gulf via amphibious operations, or both. The planners admitted, however, that such operations would pose difficult problems because of terrain and logistical considerations.[57]

Halfmoon did not describe operations beyond the war's first year. Moreover, the Chiefs recognized that Halfmoon did not sustain the national policy of supporting the nations of Western Europe and failed to provide for the initial retention of Middle East oil. Forces were simply not adequate. As in earlier plans Halfmoon relied heavily upon atomic weapons to produce decisive results, but in the absence of sufficient conventional forces the Chiefs felt they had no other alternative. On the other hand, the JCS finally had a plan that satisfied not only the individual services but also America's most important alliance partners.

In the ensuring months planners elaborated details of Halfmoon. The JSPC began work on assigning specific missions to the services and quickly became ensnarled in the roles and missions controversy. The Air Force insisted that the air offensive should be exclusively an Air Force mission. The Navy argued that carrier air should participate in strategic bombing missions.[58] The result was a series of unresolved split decisions. On the other hand, most strategic tasks were assigned without serious disputes.

In mid-June the Joint Logistics Plans Committee reported on the logistic feasibility of the first-year operations called for in Halfmoon. The JLPC noted significant shortages of aircraft and spare parts. The Committee also pointed out that there were not enough construction battalions, that the Cairo-Suez airfields were in poor repair and that there was insufficient detailed planning with the Allies. Nevertheless, the Committee believed that the plan was barely feasible and that problems and shortages could be corrected within current budget and force levels.[59] Another report by the JLPC in August noted the need for austere levels of support in many areas and pointed out that foreign-aid requirements might well deny American forces equipment necessary to execute Halfmoon operations.[60] The following day, the JLPC again noted that all operations would have to be planned on an austerity basis and that individual units could not expect standard scales of equipment and support. Moreover, aid to the Allies would increase considerably the logistical effort in many areas.[61]

On October 14, 1948, the JSPG submitted a revision of Halfmoon-Fleetwood to respond to shifts in the global political situation and to changes in military capabilities. JSPC 877/23 added East Germany, Czechoslovakia and North China to the list of Soviet satellites. They ceased regarding Finland as a satellite. The planners viewed Finland as well as West Germany, Austria, Italy, Greece, Iran, and South Korea as pro-Western nations that were too weak to assist the Allies. Portugal joined the ranks of active Allies as did Ceylon while India and Pakistan were viewed as neutrals, who under pressure would make their territory and economic assets available to the Western powers. The JSPG regarded the Arab world as anti-Western yet less ill disposed towards the Allies than to the USSR. In case of war the Arab states and Israel would reluctantly allow the use of their lands and economic assets by the Allies.[62]

The rest of the plan followed the original with minor exceptions. The Strategic Air Command would establish bases not only in England but also in Ireland. In the Middle East Aden replaced Khartoum as part of the Cairo-Suez base complex. The Allies would mobilize twenty-four divisions for offensive operations by D+12 months but only 675 aircraft.[63]

American, British and Canadian planners met in Washington from October 18 to 26, 1948. They wrote a plan, ABC101, which incorporated the basic assumptions and strategy of Halfmoon. A logistics study completed on November 12 noted that as with the American plan there were serious shortages in personnel and equipment but that the plan was feasible.[64] The Allied forces would, however, have to operate on very austere levels of equipment and even employ obsolescent types of aircraft.[65]

While working on Halfmoon, JCS planners also wrote Plan Cogwheel as a basis for determining mobilization requirements beyond the first year of the conflict. Using the same strategic assumptions as Halfmoon, the JSPC concluded that by D+12 months the Soviets would have: overrun Europe to the Pyrenees, commenced a bombing offensive against the United Kingdom, taken Turkey, north Syria, north Iraq, Iran, Kuwait, Bahrein, Korea, north China and Hokkaido, and begun a submarine offensive from Norwegian and Atlantic ports.[66] The Allies would hold Britain, the western Mediterranean up to Algiers, Egypt, Palestine, Lebanon and southern Syria. The air offensive would have destroyed twenty major urban areas in the USSR, doing great damage to Soviet industry. Additionally, the Allies by D+12 would have destroyed 85% of Russia's oil production.[67]

During the war's second year the Allies would continue their air offensive and mobilize additional forces either for the liberation of Western Europe or for an invasion of Russia via the Dardenelles. The Army by D+36 months would grow to eighty divisions. The Navy would deploy twenty-seven carriers and the Air Force would consist of 186 groups.

It soon became apparent, however, that the JCS had been unduly optimistic about the feasibility of their plans for a combined total of 27,524 naval and 38,573 Air Force planes.[68] On August 27 a logistics study raised serious questions concerning Cogwheel's feasibility after the first year of hostilities. The report found that engineer construction units were insufficient. By D+6 months the Army and Air Force needed seventy-seven construction battalions, but there would be only twenty-three, and only fifteen of them would be fully trained and equipped.[69] The shortage of construction battalions would in turn drastically reduce the number of Air Force groups operating out of Cairo-Suez-Aden. Moreover, because of budget limitations, the armed forces could not increase the number of units ready on D-Day.[70]

Other problems included a serious shortage of aircraft between D+8 months and D+18 months. Current production was insufficient to replace losses and supply aircraft for new squadrons. Specialists, including radio and radar technicians, were short in all services, and a shipping shortage necessitated a major reduction in the number of tanks and trucks supplied to fighting units.[71]

The Munitions Board on February 18, 1949, reported to the JCS that under the Cogwheel Plan it was not possible to meet aircraft production requirements. There would in fact be a 40% shortfall in airframe production and a 50% shortage of overall production because of other problems in the production of critical aircraft components.[72] If it was not possible to support the war plans after D+12 months, the initial emergency plan was useless and dangerous since it would place Allied forces in vulnerable and unsustainable positions. The service chiefs accepted these conclusions and called for a change in the plan.[73] On April 27, 1949, the JCS directed the JSPC to revise Cogwheel in light of real military capabilities.[74] The revised plan would be called Straightedge.

Before Straightedge could be completed the Munitions Board submitted another analysis of Cogwheel to the JCS on July 9, 1949. The Board stated that manpower and munitions requirements for later stages of hostilities exceeded resources and would do so for the foreseeable future. Consequently, the rates of mobilization had to be substantially slowed up to D+9 months. By M+12 manpower requirements would far exceed supply, and the continued increase in the size of the armed forces would jeopardize the supporting industrial effort of the civilian economy. Construction requirements were also far in excess of capabilities, and many raw material requirements outran supply. For example, between D+12 and D+24 industry required 9,981 million pounds of copper but only 6,338 million were available.[75]

After additional months of discussion, the Chiefs agreed that Cogwheel would be replaced by Straightedge, but that the military would continue to use Cogwheel as an interim plan.[76]

Many of the logistics problems of Cogwheel also plagued Plan Trojan, an updated version of Halfmoon-Fleetwood. Completed on January 28, 1949, Trojan dealt with the first year of a global war. It differed from Halfmoon-Fleetwood primarily by the addition of an annex that outlined an expanded atomic offensive. The atomic annex was written by the Air Force in December, 1948, and was in fact completed before the general plan. The Air Force intended to hit seventy Soviet cities with 133 atomic bombs in order to reduce substantially Russian petroleum and war industries.[77] Other minor variations included adding Italy, Greece, Ireland, Iceland, and the Philippines to the list of the Allies and the assignment of American forces to Malta, Gibraltar, and North Africa to guard the Mediterranean communications line.[78]

The JLPC informed the JCS that Trojan was not presently feasible without substantial loss of combat power and effectiveness because of shortages of specialized personnel, construction battalions, and long lead-time supply items. Furthermore, the airfields in the Cairo-Suez-Aden area were in poor shape.[79] Of the various deficiencies the lack of construction units was the most serious. Shortages ran from 60 to 70 percent of requirements, and budget limitations prevented an expansion of construction units. Even if the budget problem were solved and immediate action were taken, construction battalions would not be ready in time to meet a war crisis in July 1949.[80] Finally, if budget limitations for fiscal year 1950 remained firm, the JCS would have no choice but to revise its plans to make them conform to capabilities.[81]

General Vandenberg, the Air Force Chief of Staff, also noted that it would be inadvisable to prepare to implement a concept that was beyond U.S. and Allied capabilities. The 1950 budget would force a reduction in forces and render many operations in Trojan unfeasible. Moreover, General Vandenberg noted that Trojan lacked a worthwhile objective after the completion of the air offensive, that air defenses in England were not adequate, that logistic problems seriously reduced operational flexibility, and that the objective of regaining Middle East oil was not possible. The general then recommended that Trojan be adopted only as an interim measure and that the Joint Strategic Plans Committee prepare a new plan geared to forces that would be available under the fiscal year 1950 budget.[82]

General Bradley noted in April that the Army suffered from serious deficiencies in supply and personnel. Budget limits prevented the Army from correcting many of its problems. Finally, General Bradley recommended the creation of a new plan based on actual available human and material resources.[83] Admiral Denfeld, Chief of

Naval Operations, reported that his service lacked the men and equipment to fulfill its responsibilities in the war plan.[84] In June, General Vandenberg bluntly stated that the Air Force was unable with the means available to it to correct major deficiencies in the Trojan Plan. The shortage of Army construction battalions was the most serious problem but there were also serious shortages of key technical personnel. The Air Force, for example, needed 10,000 radar maintenance men but had only 2,200.[85] The Chief of Naval Operations noted that unless logistical problems were solved the U.S. military would be unable to execute the war plan.[86] General Bradley agreed with the view that logistics shortages and budget problems endangered the feasibility of the emergency war plan.[87]

The related problems of logistic shortfalls and a limited budget forced the JCS to abandon Cogwheel and Trojan. New plans had to be devised to conform to the force levels that would be possible under the FY1950 budget. At the same time events in Western Europe propelled the JCS in the direction of altering its strategic assumptions and plans. The creation of the Western European Union and the founding of NATO led the JCS to reexamine the possibilities of defending Western Europe.

NOTES

1. JSPG 499/2, Charioteer, 3 December 1947.
2. *Ibid.*
3. *Ibid.*
4. *Ibid.*
5. *Ibid.*
6. *Ibid.*
7. JCS 1725/14, 23 January 1948.
8. *Ibid.*
9. Munitions Board Report, 18 February 1948.
10. JSPG 500/2, Bushwacker, 8 March 1948.
11. *Ibid.*, Annex A to Appendix B.
12. *Ibid.*
13. *Ibid.*
14. *Ibid.*

15. *Ibid.*, Appendix A and Annex A to Appendix A.
16. *Ibid.*, Annex A to Appendix B.
17. *Ibid.*
18. *Ibid.*, Appendix A, Part IV.
19. *Ibid.*
20. *Ibid.*
21. *Ibid.*, Annex A to Appendix A.
22. *Ibid.*
23. *Ibid.*
24. *Ibid.*, Annex B to Appendix B.
25. *Ibid.*
26. *Ibid.*
27. *Ibid.*
28. *Ibid.*, Annex C to Appendix B.
29. *Ibid.*, Annex A to Appendix A.
30. *Ibid.*, Annex A to Appendix A, Tab A to Annex B of Appendix B and Annex C to Appendix B. Other figures in the same plan indicate a call for 180 groups. See Tab A to Annex A to Appendix A.
31. *Ibid.*, Annex A to Appendix A.
32. JSPG 496/10, Crankshaft, 11 May 1948.
33. *Ibid.*, Part V.
34. *Ibid.*, Tabs A, B and C to Annex A to Soviet force levels might be added nearly two million Chinese communist troops.
35. *Ibid.*, Annex A to Appendix.
36. *Ibid.*
37. *Ibid.*
38. *Ibid.*
39. *Ibid.*, Annex B to Appendix.
40. *Ibid.*
41. *Ibid.*
42. *Ibid.*
43. Kenneth W. Condit. *The History of the Joint Chiefs of Staff, The Joint Chiefs of Staff and National Policy, Volume II, 1947–1950.* Wilmington, Delaware: Michael Glazier, Inc., 1979, p. 288.
44. JCS 1844/4, Brief of Short-Range Emergency Plan "Halfmoon," 6 May 1948. (Halfmoon was renamed "Fleetwood" and then "Doublestar.")
45. *Ibid.*

46. *Ibid.*
47. *Ibid.*
48. *Ibid.*
49. *Ibid.*, IX Allied Plan of Action.
50. *Ibid.*
51. *Ibid.*
52. *Ibid.*
53. *Ibid.*
54. *Ibid.*
55. *Ibid.*
56. *Ibid.*
57. *Ibid.*
58. JSPC 877/3, Directives for the Implementation of "Halfmoon," 3 May 1948, and JCS 1844/7, Directives for the Implementation of Halfmoon, 26 May 1948.
59. JLPC 416/12, The Logistic Feasibility of Operations Planned, Halfmoon, 15 June 1948.
60. JCS 1844/15, Report by the Joint Logistics Plans Committee to the Joint Chiefs of Staff on the Logistic Feasibility of Halfmoon, 12 August 1948.
61. JCS 1844/15, Report by the Joint Logistics Plans Committee to the JCS, 13 August 1948.
62. JSPC 877/23, Revised Brief of Short-Range Emergency Plan Fleetwood, 14 October 1948.
63. *Ibid.*
64. JLPC 416/32, The Logistic Feasibility of ABClOl.
65. *Ibid.*
66. JCS 1725/22, Cogwheel Joint Outline War Plans for Determination of Mobilization Requirements for War Beginning 1 July 1949, 26 August 1948.
67. *Ibid.*
68. *Ibid.*
69. JCS 1725/23, Logistic Feasibility of Cogwheel, 27 August 1948.
70. *Ibid.*
71. *Ibid.*
72. JCS 1725/34, 18 February 1949.
73. JCS 1725/35, 25 February 1949 and JCS 1725/36, 28 February 1949.
74. JCS 1725/42, 27 April 1949.

75. JCS 1725/45, 9 July 1949.
76. JCS 1725/48, 9 August 1949; JCS 1725/64, 8 November 1949;
     JCS 1725/67, 1 December 1949; JCS 1725/68, 9 December
     1949; and Condit, *op. cit.*, pp. 308–309.
77. Condit, *op. cit.*, p. 293.
78. *Ibid.*,p.294.
79. JCS 1844/33, Report by the Joint Logistics Plans Committee on
     . . . the Joint Outline Emergency Plan, 6 January 1949.
80. *Ibid.*
81. *Ibid.*
82. JCS 1844/34, Memorandum by the Chief of Staff, U.S. Air
     Force, 18 January 1949.
83. JCS 1844/36, Memorandum by the Chief of Staff, U.S. Army,
     27 April 1949.
84. JCS 1844/39, 13 May 1949.
85. JCS 1844/41, Memorandum by the Chief of Staff, U.S. Air
     Force, 18 June 1949.
86. JCS 1844/42, Memorandum by the Chief of Naval Operations, 11
     July 1949.
87. JCS 1844/44, Memorandum by the Chief of Staff, U.S. Army,
     15 July 1949.

CHAPTER V

From Offtackle *to* Dropshot

By late 1948 the Joint Chiefs of Staff faced the dismal prospects of limited military resources, a declining budget, and a growing Soviet military threat. By mid-1949 the United States had 1,591,232 men under arms. The Army fielded ten divisions and a number of independent regimental combat teams. Only one of the divisions was ready for battle. The Navy had 331 combat ships, and the Air Force consisted of forty-eight groups. President Truman established the fiscal year 1950 defense budget at 14.4 billion dollars. The Chiefs protested, asserting that the budget would force major reductions in conventional forces and render impossible crucial operations in Europe and the Middle East.[1] Truman, however, remained firm, and the future held little prospect of increased defense spending.

While the JCS viewed American power as stagnant or even declining, they believed that Soviet capabilities were steadily growing. On February 16, 1948, the Joint Intelligence Committee presented the JCS with a study of the expansion of Russian military power during the forthcoming decade. In 1948 the Soviets could overrun Europe, the Middle East, North China, and northern Japan, but Moscow was not yet able to attack directly the Western Hemisphere. By 1957, however, the USSR would possess atomic bombs and a long-range strategic air force. In case of war the Soviets would simultaneously execute offensive operations in Eurasia and seek to cripple the United States by mounting atomic and CBR attacks coupled with extensive sabotage and subversive activities. The air threat to the United Kingdom would also grow progressively more serious as would the power of the Soviet submarine fleet.[2]

Additional long-range forecasts in the summer of 1948 indicated that by 1957 the USSR would have an army of 175 divisions, 15,000 combat aircraft and 750 submarines.[3] Intelligence analysts presumed that a Soviet rifle division was equal to a British or French infantry division and had half the combat power of an American infantry division. Soviet tank and mechanized divisions were equal to similar British and French units and had two-thirds the combat power of American armored divisions.[4]

A more detailed threat appraisal appeared on November 30, 1948. The Joint Intelligence Committee noted that as of August 1, 1948, the Soviet army contained 2.5 million troops. The order of battle included 104 rifle, thirty-five mechanized, ten tank and fifteen cavalry divisions. Thirty-one Soviet divisions were stationed in Eastern Europe. There also existed ninety satellite divisions, the best of which were the thirty divisions of the Yugoslav army. Upon mobilization the Red Army would expand quickly. By M+5 days all of the divisions in being would reach full wartime strength. By M+30 another 125 to 145 divisions would be ready to take the field, and by M+420 the Red Army would have 500 line divisions.[5]

The Soviet Air Force consisted of 500,000 men and 15,000 aircraft. By M+6 months an additional 5,000 combat aircraft would be operational. Moreover, the USSR was producing German model V1- and V2-type missiles. After 1950 the Russians would possess a growing supply of atomic bombs and a substantial capability to wage chemical and biological warfare. By 1956–1957 the Red Air Force would be able to launch air strikes directly against the U.S. and Canada.[6]

The Soviet Navy numbered 600,000 men including 275,000 coast defense troops. The order of battle included 110 coastal defense submarines and 163 long-range U-boats. Surface strength was marginal and unlikely to pose a serious threat currently or in the future although long-range submarine strength would expand.[7]

The Russians had the forces to launch offensives in Europe, the Middle East, China, and Korea simultaneously. Logistical support was sufficient to support all of the operations. By 1956–1957 air and naval forces would be larger. The Soviets would have between twenty and fifty A-bombs and a rocket inventory with conventional, biological, and chemical warheads. The army would not grow numerically but would receive more modern equipment, and many rifle divisions would evolve into tank or mechanized units.[8]

The JCS response to the Intelligence Committee estimates was to examine their ramifications and occasionally recommend increased American capabilities. A special report by the Research and Development Board on biological warfare noted that biological warfare techniques in the USSR were well advanced. The United States, therefore, had to maintain and improve offensive and defensive capabilities. Submitted on February 20, 1948, the JCS approved the report in late May.[9]

The JCS also became increasingly alarmed by the threats of sabotage and subversion. On March 22, 1948, the Joint Intelligence Group presented a long-range estimate of communist influence in the United States. The report stated that Negroes, people of recent European origin, professionals, youth and women's organizations, architects, and writers were ripe for infiltration. Persons of old American stock, especially liberals concerned with peace and social justice, were also highly susceptible to communist propaganda. Over all, thirty-five percent of the population was in danger of being infiltrated and subverted. The JIG predicted that the communist network would continue to expand and in wartime could cause major disruptions of the national mobilization effort. Furthermore, saboteurs using atomic devices and bacteriological weapons could inflict significant losses on the population.[10]

A report on current subversive efforts, submitted in mid-September, 1948, placed membership in the Communist Party at 68,500. There were also 685,000 fellow travelers. Negroes and the foreign born were highly susceptible to Soviet propaganda as was Henry Wallace's Progressive Party. The number and influence of subversives would continue to grow and would pose a significant threat in time of war.[11]

The extent of Soviet military power forced U.S. planners to reexamine their assumptions concerning the importance of retaining or recapturing at an early date Middle Eastern oil resources. Faced with the inability to keep Soviet armies at bay in the region, diplomatic and military analysts decided that the region's oil was not in fact absolutely necessary for the Anglo-American war effort and began to draw up plans to destroy the wells in order to deny them to the Russians.

On May 6 and again on May 25, 1948, the State Army Navy Air Force Coordinating Committee, SANACC, produced papers on the destruction of Middle East oil wells. The papers recommended plugging the wells. A plugged well could never again produce oil, and oil could be recovered only by digging new wells. The United States would assume responsibility for destroying Saudi wells, the British would plug Iranian wells, and the demolition of wells in Bahrein would be a joint Anglo-American venture.[12]

SANACC suggested that demolition plans be readied well in advance of hostilities. An officer, operating under a commercial cover, should organize secret agents, Allied diplomatic and military missions, and oil company personnel to carry out the demolitions. Host countries would not be informed of the plans or the existence of the secret sabotage networks.[13] Thus, perceptions of overwhelming Soviet military power doubtless coupled with declining British influence in the region led American planners to modify drastically their view of the region's strategic importance.

Austere budgets, increasing Soviet military power and the knowledge that the USSR was diligently working to develop atomic weapons convinced the JCS of the need to expand its own atomic arsenal. In December, 1947, the JCS noted that the Air Force had thirty-three Silver Plate aircraft and that by November, 1948, there would exist 120 bombers capable of delivering atomic weapons. There would also be three bomb assembly teams by June, 1948, and seven by July, 1949. The JCS also called for the production of 400 atomic bombs by January 1, 1953.[14] In March, 1948, the JIC noted that the Soviets would test their first atomic weapon sometime between 1950 and 1953 and by 1953 would possess a stockpile of twenty to fifty atomic bombs, depending upon when they tested their first atomic device.[15]

On April 17, 1948, the JCS informed the AEC that atomic weapon stockpiles had to be ready before the start of war in order to sustain an immediate air offensive.[16] In July, 1948, the JCS requested an interim figure of 150 bombs to hit 100 urban targets, and in May, 1949, the Air Force Chief of Staff called for sufficient bombs to strike 220 targets.[17] The JCS had, meanwhile, advocated a substantial increase in the 1953 stockpile.[18] A special report, however, called into question a number of assumptions of American atomic strategy.

The JCS appointed a special committee, headed by Lieutenant General Hubert R. Harmon, United States Air Force, to respond to questions about the effectiveness of atomic attacks raised by Secretary of Defense Forrestal. Composed of two senior officers from each service, the Harmon Committee examined the effects of an atomic offensive against the USSR through May, 1949. The Committee presented a unanimous report to the JCS on May 12, 1949.

The Committee assumed that the Strategic Air Command attacked seventy Soviet cities and was completely successful in delivering all of its weapons to the appropriate targets. The initial attacks would kill 2.7 million people, wound an additional 4 million and make life very difficult for the remaining 28 million residents of the target cities. The attacks would also cause an immediate 30 to 40 percent reduction of Soviet industrial capacity.[19]

A successful atomic offensive would not, however, force the USSR to end hostilities on American terms. Soviet industry would recuperate quickly although additional bombing could do great damage to the petroleum industry. Moreover, the Russian armed forces had sufficient supplies to execute their initial offensive operations despite dislocations on the home front. Finally, the report stated that an atomic offensive, far from undermining Soviet morale and weakening Moscow's hold on the populace, would stimulate resentment against the United States, validate Soviet propaganda against foreign nations, unify the public and strengthen their will to fight.[20]

The Harmon Committee did not, however, reject the idea of an atomic offensive and noted that given the state of America's armed forces, atomic weapons were the only means of rapidly striking at the USSR's war-making capacity.[21] The absence of effective conventional alternatives left the United States with no choice but to rely heavily on atomic weapons. In mid-May, 1949, an *ad hoc* JCS committee recommended a vast expansion of the projected atomic bomb stockpile. The following month the JCS forwarded the report to the Atomic Energy Commission. The new arsenal was supposed to be about three times as large as the original goal.[22]

The discovery in the summer of 1949 that the Soviets had exploded an atomic device did not much disturb the military establishment. They expected the Russians to develop an atomic arsenal perhaps as early as

1950, and several war plans had already designated the USSR's atomic production and storage facilities as primary targets.

On November 16, 1949, General Vandenberg, the Air Force Chief of Staff, noted that in a few years the Russians would possess a large number of atomic weapons and could with fifty or sixty weapons devastate the United States. His solution was not more A-bombs since the U.S. would have a plentiful supply by 1953 but rather improved air defenses.[23] At the end of January, 1950, the JCS estimated that by mid-1950 the USSR would have from ten to twenty atomic bombs and that the stock would rise to between seventy and 135 by mid-1953. The U.S. for its defense needed thirty-seven new squadrons, bringing the total to sixty-seven air defense squadrons.[24]

Thus, throughout 1949, JCS planners were driven to increase their reliance upon atomic weapons. Budgets and force structures seemed to lead the JCS inevitably to fewer conventional force commitments in Europe and the Middle East and to increasing the numbers of atomic bombs and targets. Diplomatic events in Europe, however, soon began to exercise a countervailing influence, forcing the Chiefs of Staff to contemplate what they had hitherto regarded as impossible--the conventional defense of Western Europe.

Signed on March 17, 1948, the Brussels Treaty created the Western European Union, a defensive alliance which included England, France, Belgium, Holland, and Luxemburg. Washington encouraged the European powers to unite in a collective defense effort, and the JCS quickly recognized that it had to respond to this new political imperative.

On April 17, 1948, the Joint Strategic Survey Committee submitted a report on the military implications of the Brussels Pact. The JSSC recognized that the American government favored European defense efforts. The military, therefore, had to encourage their efforts by committing its forces to the defense of Western Europe. Without such a commitment the Europeans would be psychologically and militarily unable to resist Soviet pressure. The JSSC concluded that such a commitment required expanded U.S. forces. Current force levels were barely adequate to execute current emergency war plans. Any attempt to try to defend Western Europe with current forces would dangerously over-extend the American military.[25]

The Joint Strategic Plans Committee submitted a more detailed analysis of the military implications of the Brussels Pact on April 21,

1948. The JSPC asserted that the Western European nations needed a U.S. commitment to resist overt Soviet aggression in order to develop the confidence to pursue economic recovery programs. The Czechs did not resist the Communist coup because of the massive Soviet military presence on their borders. To avoid future Czech-style coups the United States had to fill the military vacuum in Western Europe. An American commitment to defend Western Europe would enable states in the region to resist both internal communist pressure and external Soviet military threats.[26]

If the United States undertook the defense of Western Europe, the JSPC pointed out that war plans would have to be changed to reflect the new political situation. Emergency war plans would require an effort to hold a lodgement area on the Continent, and if Allied forces were driven off the Continent entirely, the first call on reinforcements would be the rapid liberation of Western Europe. Medium-range plans would have to reflect an effort to defend the Rhine and parts of the eastern Mediterranean. The Air Force as its first priority would assist in halting the advance of the Soviet armies and only then would it execute offensive operations against the USSR's war industries.[27]

American officers began to attend Western European Union meetings on an informal basis, and in July, the U.S. agreed to a WEU emergency war plan that called for a defense of the Rhine and coordinated the actions of American, British, and French forces.[28] On September 22, 1948, the JCS approved a command plan for Halfmoon-Fleetwood which provided an Anglo-American Combined Chiefs of Staff. The plan also designated a British or French officer to command Allied forces in Western Europe and a British officer to act as Allied Commander-in-Chief in the Middle East.[29] In March, 1949, the JCS approved a short-term WEU defense plan since it was in accord with current U.S. plans.[30]

The WEU plan called for holding a Soviet offensive as far to the east as possible. The Rhine was probably the major defense line in Europe. The Allies would also hold the Middle East as an offensive base and North Africa. The Dutch asked that their northern provinces be included within the defensive perimeter, and the other powers agreed.[31]

By the summer of 1949 the WEU's short-term defense plan called for holding the Rhine-Ijssel line in central Europe. The Allies also intended to hold positions from the Italian Alps south to the Adriatic and to assist as far as possible Danish and Norwegian defense efforts.[32] The American military believed the plans were reasonable especially in the long term. U.S. aid could strengthen European forces, but there was a serious short-term risk. Aiding European forces would deprive American forces of vital material and could restrict U.S. self-defense efforts for several years. Support of the WEU was thus a gamble that posed immediate risks but promised long-term strategic benefits.[33]

The creation of the North Atlantic Treaty Organization essentially transformed informal military relationships into formal commitments. Many WEU institutions became NATO regional organizations, and Brussels Pact strategy was readily applicable to plans to defend the North Atlantic Pact. The formal American commitment to defend Western Europe did, however, place JCS planners in an ironic situation. Fiscal austerity and force limitations indicated that American strategy had to emphasize the employment of atomic weapons and limit reliance on conventional operations. Political commitments on the other hand seemed to require serious efforts to provide conventional defense for Western Europe.

General Dwight D. Eisenhower left private life to serve as a special advisor to the Secretary of Defense and to act as an unofficial presiding officer of the JCS from February to August, 1949. Eisenhower was primarily concerned with the FY1951 defense budget, but he also recognized that despite the constraints imposed by austere defense appropriations the growing American involvement with Europe required revisions of American strategic concepts. He, therefore, called for a new emergency war plan that included the defense of Western Europe. If it were not possible to hold the Rhine, Eisenhower called for the retention of a continental bridgehead, and failing this for the reinvasion and liberation of Europe as soon as possible.[34]

On April 26, 1949, the JCS sent to the Joint Strategic Plans Committee guidance for preparing an emergency war plan. The plan would cover the first two years of a conflict beginning on 1 July 1949, and force levels would be those available under the fiscal year 1950 budget.[35]

The service chiefs told the planners that the overall concept would involve a strategic offensive in Western Eurasia and a strategic defensive in the Far East.[36] They directed that basic undertakings include the protection of the integrity of the Western Hemisphere including the defense of the continental United States, the Fairbanks-Anchorage-

Kodiak area of Alaska, Venezuelian oil areas and the Panama Canal.[37]
Other essential defensive tasks included the protection of Iceland,
Greenland, the Azores, Okinawa, and Japan. American and Allied forces
would also secure vital sea lines of communication, defend England
against invasion and disabling air attack, secure North Africa and the
Cairo-Suez area, and launch as soon as possible a strategic air offensive
against Soviet war industries.[38]

The JCS considered that American security required the
development with European allies of a strategy designed to cover
Western Europe no farther to the west than the Rhine. Pending the
attainment of forces sufficient to hold a line from the United Kingdom-
Rhine-Cairo-Suez, the JCS called for plans to hold a substantial
bridgehead in Western Europe. If it were not feasible to hold a
bridgehead, the Chiefs told the planners that war plans should envisage
a return to the Continent at the earliest practicable time.[39]

The Joint Strategic Plans Committee submitted its new operational
concept to the JCS on November 8, 1949. Plan Offtackle demonstrated
the difficulties inherent in attempting to respond to national policy
objectives with inadequate forces. The opening sections of Offtackle
were in fact a discussion of the calculated risks arising from force
limitations. The planners pointed out that ground forces deployed
during the first year of hostilities would not have the full combat
equipment specified in current tables of organization. Moreover, if
divisions were committed to battle at an early date, additional units
deployed after D+6 months would suffer a considerable loss of combat
effectiveness because of cannibalization of manpower and essential
material to sustain initial deployments.[40]

There also existed a shortage of aircraft and spare parts which would
reduce operational sortie rates. All services suffered shortages of
technical and specialist personnel, supply items, and construction units.
Finally, in the early months of the war, there would be a serious
shortage of aviation fuels.[41] The planners, nevertheless, felt that they
had to accept these shortcomings if they were to produce a document
that fulfilled national policy goals.

Offtackle was the first plan that contained authoritative political
guidance. NSC 20/4 established American objectives toward the
USSR. Political goals in both peace and war were identical--to reduce
the power and influence of the USSR to limits which no longer

constituted a threat to peace and the independence and stability of the world family of nations.[42]

War aims supplemental to peacetime goals included the elimination of Russian domination in areas outside the borders of any Russian state allowed to exist after the war and the destruction of the network of relationships through which the Communist Party of the USSR exerted influence over groups and individuals in the non-communist world.[43]

The United States also intended that any regime or regimes on traditional Russian territory after the war would not have the requisite military power to wage aggressive war. Additionally, if any Bolshevik regime survived in any part of the USSR, it would be denied the military-industrial potential to be able to wage war against any other regime or regimes existent on traditional Russian territory. These goals were to be pursued without permanently impairing the American economy or way of life.[44]

NSC 20/4, like earlier statements of political objectives in JCS plans, did not call for unconditional surrender but nothing short of complete capitulation or collapse would permit their realization. NSC 20/4, in contrast to earlier military plans, did clearly envision the breakup of the Soviet state and the emergence of several political entities to replace the USSR. How these goals could be achieved without a massive American effort remained unanswered. The military planners were, however, satisfied, for at least they could be sure that the politicians viewed a conflict with the USSR much as they did. And war with the Soviets would be global and total and would be pursued as long as necessary to achieve the nation's political aims. War would be fought with virtually unrestricted intensity and there would be no substitute for victory.

Military planners also made a number of predictions and assumptions about a war against the USSR. China would by 1950 be under Communist control except for Hong Kong, Formosa, Macao, and possibly some remote areas in the hinterland. Mao would assist the Russians and follow a policy of opportunist expansion. On the other hand, Yugoslavia would refuse to join the USSR in a war against the West.[45]

America for its part could count upon the NATO nations, Australia, New Zealand, South Africa, and Ceylon. India and Pakistan would remain neutral. The South American republics, the Philippines, and Japan would cooperate with the Allies while West Germany,

Austria, Greece, Turkey, Iran, Iraq, South Korea, and Indo-China would be sympathetic although their strategic and political position might be so difficult as to prevent some of them from active cooperation. The Arab states would make their economic resources and territories available to the Allies. Israel would also assist the Allies if threatened by the USSR.[46]

Additional planning assumptions included the view that Western European nations would be unable to halt a major Soviet attack and that atomic weapons would be used by both sides. By the end of 1950 the USSR would have about thirty atom bombs. Considerations of retaliation might lead both sides to refrain from employing chemical and bacteriological weapons. Finally, the planners in contrast to earlier concerns for Middle Eastern oil resources stated that the oil of the region was not vital to the Allied war effort. The JSPC offered no explanation for the sudden reversal and noted that the reconquest of oil-producing areas would be undertaken during the second year of hostilities only if operations in the Gulf region did not detract from more critical tasks.[47]

The Soviets would start the war either as a result of miscalculation or because Moscow decided to act in order to forestall further Western European political, economic, and military recovery. The war would begin with little or no warning. Consequently, Mobilization Day and D-Day would be virtually identical.[48] Since wars are almost invariably preceded by political if not tactical warning, the JSPC was producing a plan predicated on a worst-case analysis.

The planners believed that the USSR had armed forces adequate to undertake simultaneous attacks in Western Europe, the Middle East and Far East, and also retain an adequate reserve. The Soviets could also launch attacks with limited objectives against Canada and Alaska, mount a subversion and sabotage campaign against Allied interests throughout the world, bomb the British Isles and attack Allied sea lines of communication. After the conclusion of their initial ground offensives, the USSR would be able to mount a full-scale air offensive against Britain and, if necessary, attack Spain and Scandinavia. The Russians could also attack Allied air bases, if any, in Pakistan and mount diversionary operations to cause the Allies to maldeploy their forces.[49]

The JSPC did not explain how the USSR would attack Italy without Yugoslav support. Would Belgrade cooperate with Moscow or resist the passage of Soviet troops? If Tito resisted, would the Allies give him military assistance? Would the Russians simply by-pass Yugoslavia and try to invade Italy via the Alps? The strategic implications of the Moscow-Belgrade split were simply ignored as were the results of the Communist victory in China. Would Mao attack India or Indo-China? What would the Russian forces in the Far East do? Would they attack northern Japan or the Aleutians or would they be transferred west? The planners were silent on these strategic issues, presuming only that a war would in its initial phases be marked by large-scale, successful Russian offensives.

During the war's first phase, D to D+3 months, the Allies would secure areas vital to the prosecution of hostilities and launch an atomic air offensive against the USSR. The United States and Canada would activate a limited air-warning net and fighter groups for the defense of critical industrial areas. Due to manpower and equipment shortages air defense forces would have to consist of units either scheduled for overseas deployment or engaged in training. Two regiments would guard atomic installations and stockpiles of atom bombs against sabotage; a regiment would secure the Sault St. Marie Canal, and three division equivalents would secure other critical areas.[50] No mention was made of the impact of a Soviet atomic attack, nor was there any discussion of how the Russians would use their atomic arsenal or whether Moscow's possession of a limited number of A-bombs would compel the Americans to alter their nuclear strategy.

The planners assigned two regiments, one-and-a-third fighter groups and five antiaircraft battalions to defend Alaskan bases. Two regiments, two antiaircraft battalions and a fighter squadron would protect the Panama Canal. An airborne division, two Marine battalions with their amphibious lift, an air mobile Canadian regiment, and two Canadian fighter squadrons constituted the Western Hemisphere reserve.[51]

As soon as possible after D-Day, two Marine battalions would occupy Iceland, and another Marine battalion would proceed to the Azores. Marine fighter squadrons would provide any necessary air defense. In the Far East four-and-a-third army infantry divisions, nine antiaircraft battalions, two carriers, and five fighter groups would guard Japan, the Ryukyu, Okinawa, the Philippines, and Formosa. The Navy and Air Force would also seek to destroy the Soviet fleet, ports and air installations.[52]

In Western Europe the JSPC recognized that the Allies were not strong enough to hold the Rhine. Offtackle, therefore, contained two alternatives: holding a substantial bridgehead or withdrawing from the Continent with the intention of reinvading as soon as possible. The planners felt that in 1950 the Allies would still be too weak to hold a bridgehead. Therefore, the Allies would plan a fighting retreat either to the line of the Pyrenees or to western or southern French ports for evacuation to England or North Africa depending upon the military situation. Allied forces in Trieste and Austria would also retreat from the Continent.[53]

The JSPC believed that the Russians lacked the sea and airlift capacity to pose an invasion threat to the United Kingdom in 1949–1950. British local defense forces, three infantry and one armored division plus five independent brigades, provided adequate security against direct attack. The Royal Navy was capable of handling any submarines or surface raiders although a large-scale mining effort could pose a serious threat to Britain's sea lines of communication. It would require enormous expenditures of Allied resources to counter Soviet mine warfare operations.[54]

The air threat to Great Britain was significant. By D+2 months the Soviets would be able to deliver 12 to 16,000 tons of bombs per month plus 11,000 fighter sorties. Adequate air defense of key bases and installations required 144 antiaircraft regiments and 1,152 fighters. By D+1 month the British would have 107 AA regiments and 232 fighters. The Americans would have to supply thirteen AA battalions and two fighter groups, but even with these reinforcements, air defense forces would fall considerably short of those needed to provide adequate protection. The Allies, however, could supply no further forces unless air defense forces in North Africa and other areas were redeployed to England. If it proved absolutely necessary in order to protect the United Kingdom, the Allies would redeploy air and AA units even though such actions would jeopardize the success of other critical tasks.[55]

In the western Mediterranean the Allies would secure the North African coastline as far east as Tunisia and establish bases at Oran and Algiers. By D+3 months the United States Army would have four divisions, two regimental combat teams, and twelve antiaircraft battalions supported by one-and-a-third fighter groups in the western Mediterranean. A Marine regiment and ten fighter squadrons would also deploy to North Africa.[56] Since the planners believed that the Allies could not hold a continental bridgehead, North Africa would in later phases of the war become one of the major places from which Allied forces would reinvade the European mainland.

Following the earlier Pincher Plans, the JSPC presumed that the Soviets could overrun Spain but would not attempt to advance into Iberia in order to conserve forces for use elsewhere. If the Red Army did attack, Allied forces would try to hold them at the Pyrenees and if they failed would establish a bridgehead north of Gibraltar.[57]

In the eastern Mediterranean the British would maintain control of the Cairo-Suez area. By D+3 months the British would have three-and-two-thirds divisions and three armored car regiments in the Middle East supported by more than 100 fighters and 100 bombers. The British would carry out demolitions of the transportation network, oil wells and refineries in the Gulf Region. Logistical constraints would prevent the Russians from sending more than nine divisions into the Middle East, and it would take the Red Army 240 days to reach northern Palestine. The JSPC, therefore, presumed that the British had a good chance of holding a line north of Cairo-Suez.[58]

Seven American aircraft carriers would operate in the eastern Atlantic and Mediterranean. If necessary, the carriers would provide direct support to the defense of the United Kingdom. Otherwise, they would assist Allied ground and air forces in North Africa while other naval forces secured essential lines of communication. If Soviet forces threatened the line of communications in the Mediterranean, the Allies would establish an alternate route around the Cape of Good Hope.[59]

While the Allies carried out their defensive tasks, the Strategic Air Command would launch an atomic offensive at the earliest possible date after D-Day. By early 1949 SAC had about 120 nuclear-capable aircraft, thirty of which could be refueled while in flight. Nineteen B-29s had been transformed into aerial tankers. Six bomb assembly teams were fully trained and another was receiving instruction.[60] The Air Force would operate from bases in the United Kingdom, the continental United States, and Okinawa. Bases in Iceland, North Africa and the Middle East would also be used for staging. At the start of the air offensive three groups would operate from the United States, two from England, and one from Okinawa. By D+3 months one bomber group would remain in the US, one on Okinawa, and forces in England would expand to seven bomber, two reconnaissance, and five fighter escort groups.[61]

The air offensive would strike at vital elements of the USSR's war-making capacity and seek to retard the Red Army's advance into western Eurasia. To achieve these objectives SAC would attack 104 Soviet cities with 220 atomic bombs. SAC would also hold seventy-two A-bombs for reattack. Most of the weapons would be delivered by D+3 months. New atomic bombs and conventional bombing would maintain existing damage levels, and if necessary, the Air Force would attack additional target systems.[62]

During the war's second phase, D+3 to D+12 months, the Allies would continue the strategic air offensive, accelerate mobilization, and build-up forces in England and North Africa in preparation for a reinvasion of Western Europe. Because of prewar stockpiling Soviet armies, despite the air offensive, would still be able to continue operating, thus making reinvasion necessary. Allied forces would also continue to maintain defensive positions established in the first phase of hostilities and undertake limited offensive operations designed to keep the Soviets off balance and seize forward bases for the liberation of Europe.[63]

For the defense of the United Kingdom the United States by D+6 months would send thirty-five antiaircraft battalions, four fighter, and one light-bomber group to Great Britain. In the Cairo-Suez region British and Dominion forces by D+12 months would expand to seven infantry divisions, five armored regiments, and eighteen defense battalions supported by as many as eight American carrier battle groups. In North Africa the American build-up by D+12 months would consist of twelve divisions, seven infantry, two armored, and one airborne, two independent regiments, and forty-eight antiaircraft battalions supported by three-and-a-third fighter and two light-bomber groups. The Navy in addition to the carriers would maintain a Marine division and its amphibious lift in the region plus assault lift for four Army divisions.[64]

As the Allied build-up developed, Anglo-American forces might undertake a number of limited offensives if such operations did not materially delay the ultimate return to Western Europe. Depending upon the circumstances, the Allies would attempt to seize Sicily, Southern Italy or Sardinia and/or Corsica. Occupation of one, some, or all of these objectives would offer additional security for the Mediterranean line of communications and provide advanced bases for tactical air units.[65]

The Allies would reenter Western Europe during the war's third phase, D+12 to D+24 months. While Allied air power continued to attack Soviet industrial and military targets, forty-one American divisions and ten carrier battlegroups would prepare to launch a two-pronged invasion supported by eleven light bomber and thirty-two-and-a-third fighter groups.[66]

About two-thirds of the available forces, launched from North Africa, would land in southern France and advance up the Rhone Valley. The advance would then pivot eastward. The second offensive would jump off from England, landing at favorable locations between Cherbourg and the base of the Jutland Peninsula as far to the east as possible. Forces in the north would advance south, join forces with the units advancing from the Rhone Valley and isolate major Soviet forces in Western Europe. After the pincers closed, the Allies would attack to the east to limit Soviet opportunities to consolidate defensive positions.[67]

During the final phase of hostilities, the Allies would continue their drive eastward using fast-moving mechanized columns and airborne operations to isolate large pockets of Russian troops. If Moscow did not capitulate, the Allies, after liberating Europe, would proceed to invade the USSR. The Allies would advance until the Soviet government collapsed or surrendered.[68]

The planners did not discuss what the Allies would do if the Soviet Government, defeated in battle, resorted to guerrilla warfare and were probably over-sanguine about the prospects of a quick liberation of Europe. Offtackle, however, was never intended as a detailed plan for the entire war. Its main concern was to establish a concept of operations and deployment for the opening phases.

Offtackle responded to the American political commitment to defend Western Europe. European defense or more probably reinvasion and liberation of the Continent became the primary strategic objective of Plan Offtackle. Europe's security was no longer a by-product of Soviet defeat on other fronts--it was by late 1949 the primary focus of American strategic planning.

If Offtackle was politically appropriate, serious doubts remained concerning its military feasibility. The authors of the plan realized that they lacked adequate forces to execute it. Their view was soon confirmed by the Joint Logistics Plans Committee which had in fact been examining the logistic implications of Offtackle while the JSPC was working on it.

On November 15, 1949, the JLPC informed the JCS that Offtackle was infeasible in terms of carrier aircraft, light- and medium-bombers, and fighters. There were also serious shortages of technical personnel, construction units, and aviation fuel. Aid to the Allies would deplete American inventories and increase existing deficiencies. Many army combat units would have to go into battle at about half their normal effectiveness due to the lack of modern weapons and munitions. Finally, the JLPC noted that the required revision of mobilization plan Cogwheel would force changes in Offtackle if mobilization and operations plans were to be mutually supporting.[69]

Despite logistics problems the JCS in December, 1949, and again in late January, 1950, approved Offtackle, which would be retained as the basic emergency war plan until mid-1951.[70] On February 18, 1950, the JSPC issued directives for the plan's implementation.[71]

The JCS realized that the national commitment to defend Western Europe required a detailed examination of long-range requirements. A mid-range war plan for use in operational, mobilization, and fiscal planning was essential for the integration of American strategy with budgetary reality. American political and military leaders had to attempt to figure out how to defend Europe and how much it would cost. Consequently, even before the JSPC began to work out the details of Offtackle, the JCS on August 16, 1948, appointed an *ad hoc* committee consisting of an Army and an Air Force Major General and a Rear Admiral and ordered them to develop an outline plan for a war beginning in July 1956 (the date was later changed to January 1, 1957). The plan was to include the broad strategic concept and an estimate of minimum manpower and material requirements.[72]

On January 31, 1949, the *ad hoc* committee submitted its report, entitled Dropshot, to the JCS. The report began with a financial estimate. The Committee stated that the United States should spend about ten percent of the national income between 1950 and 1957 on the American armed forces, the European Recovery Program, and military aid to the Allies. Such expenditures, amounting to twenty to twenty-two billion dollars annually, would enable the United States and its allies to create forces of sufficient size and strength to have a reasonably good chance of defeating the USSR.[73]

The Committee assumed that the political objective of a conflict would be the same as the goals outlined in NSC 20/4. American allies would consist of Britain, the Commonwealth, except India and Pakistan, and the Philippines. Latin American countries would remain neutral or ally with the United States while Ireland, Spain, Switzerland, Sweden, Greece, Turkey, Pakistan, and the Arab League members would seek to remain neutral but would fight alongside the Allies if attacked. Most West European states would be American allies while Iran, Afghanistan, and India would submit to armed occupation rather than fight. The USSR would have as allies its East European satellites including Yugoslavia and Finland plus North Korea and Communist China.[74]

The Committee also assumed that West European Union forces would be capable of substantial coordinated military action and that atomic weapons would be employed by both the U.S. and the USSR. Radiological, biological, and chemical warfare was, subject to considerations of retaliation and effectiveness, also possible.[75]

On D-Day the Soviet Army would consist of 135 line divisions and twenty artillery divisions. The satellites would contribute an additional 115 divisions plus 1,450,000 Chinese Communist troops. By D+30 the Red Army would expand to 248 divisions, and satellite forces would number 175 divisions. By D+1 year the Red Army would include over 500 divisions while the satellites fielded 274 divisions. The Soviet fleet would consist of thirty cruisers, 260 destroyer-types, 300 to 350 submarines, and 200 to 300 coast defense vessels. The Red Air Force would consist of 5,400 fighters, 3,300 ground attack aircraft, 2,800 light- and medium-bombers, 1,600 reconnaissance types, 2,100 home defense interceptors, and 1,800 long-range bombers.[76]

The Committee estimated that Soviet morale would be able to bear the rigors of a general war. The planners recognized that certain elements of the Soviet population, particularly ethnic groups in the Baltic states, the Ukraine, the Caucasus and Central Asia were dissatisfied with the Soviet political and economic system and with Great Russian domination. The police and propaganda apparatus would, however, be able to control dissent, and Soviet morale would not become a serious factor until such time as a drastic deterioration of the USSR's military position took place. The populace of the satellite states also resented Russian domination but would not form effective resistance movements during the early stages of a war. Only the hope of liberation plus concrete aid from the West could transform passive hostility into active resistance.[77] Thus, morale would not affect Soviet battlefield or home-front performances until the tide of battle had decisively turned.

Although the USSR's economy would in 1957 be only about half the size of the West's, it was judged more than adequate to sustain a major war. Oil refining capacity and the transportation system would be the major shortcoming of the military economy and logistics system. Nevertheless, the Committee believed that the Soviets could by use of rail, water, and road transport sustain their offensive operations during the opening phase of hostilities.[78]

Moscow's goal in a war would be the defeat of the United States and the attainment of world domination. The military strategy would involve an invasion of Western Europe followed by an air and sea campaign to neutralize the British Isles. Simultaneous campaigns would seek to overrun Italy and Sicily and close the eastern Mediterranean while other forces advanced into Scandinavia and the Middle East. The Russians would also mount air attacks on the United States and Canada, and subversive and sabotage campaigns against Anglo-American interests in all parts of the world.[79]

Defense of the Western Hemisphere and protection of America's war-making potential had to receive first priority in all considerations of Allied strategy. Defenses had to be in place by D-Day and would consist of five-and-two-thirds divisions, 1,890 all-weather fighters, 342 penetration fighters, 108 reconnaissance aircraft, 325 antiaircraft gun battalions, 64-1/4 guided missile antiaircraft squadrons, 163 ground-control intercept stations, and 199 air warning battalions. Additional units would occupy Iceland and the Azores as soon after D-Day as possible. A regiment and a fighter group would garrison Iceland, and a battalion and a fighter squadron would hold the Azores.[80]

The Allies would also seek to hold the United Kingdom and Western Europe as far to the east as possible. The best defensive position was the Rhine-Alps-Piave line, for although it gave up the Ruhr, its successful defense would preserve the bulk of Europe's industrial and population resources. The line also presented formidable natural barriers to attacking forces.[81]

The defense of the Rhine-Alps-Piave line required seventy-six divisions and 4,500 aircraft. The vast majority of the combat forces would be Allied units, with the Americans supplying material, air power, and those ground forces currently stationed in Europe. Allied forces, however, numbered only thirty-five divisions, and a major U.S. military aid program was necessary to enable the Allies to hold the defensive position. The Rhine-Alps-Piave line was the best alternative, but the planning committee recognized that if the aid program failed to strengthen the Allies in time to counter a Soviet attack, the Western nations would require additional strategic options.[82]

The Allies could make a stand on the Rhine-French-Italian border line, a position that could be held with seventy divisions and 4,400 aircraft. The Rhine-French-Italian border position would give up Italy but still protect England and most of Western Europe. Large-scale American aid was also required to make Allied defense efforts feasible.[83]

The planners next studied the prospects for holding bridgeheads in France. Ten divisions could hold the Cotentin Peninsula, and twenty would suffice to secure Brittany. Limited port and airfield facilities, however, would probably doom such efforts, and the planners instead recommended that the Allies' third alternative should be a defense of the Pyrenees and the southern portion of Italy. Thirty-four divisions and 800 aircraft would suffice to defend the Pyrenees while twenty divisions and 500 aircraft could hold a line in Italy stretching from Rimini to La Spezia. Neither area could be developed into a major base region for offensive operations, but the retention of Spain and part of Italy would be of material assistance to Allied efforts to hold the Mediterranean and Middle East.[84]

If Italy were overrun, the Allies would then seek to stand at the Pyrenees. The final alternative was to hold Malta, Crete and Cyprus, a relatively easy task that could be accomplished with small forces. Holding the islands would do little for the defense or reconquest of Europe but would contribute to sustaining the Allied position in the Middle East.[85]

The planners believed that the Allies should hold as much of the Middle East as possible. Cairo-Suez was a critical strategic base area, and the region's oil was vital to the Allied war effort. Holding all of Turkey was not feasible because of the excessive logistics effort involved. Rather, the Committee recommended that seven Allied divisions and six fighter groups plus eleven Turkish divisions hold the southern part of Anatolia plus a line southward to Jordan. The Allies would also attempt to hold the oil producing regions of the Persian

Gulf or, if this proved to be impossible, to retake Bahrein, Basra, Kuwait and Abadan as soon as possible.[86]

In the Far East the Allies would fight a defensive war, holding Okinawa and Japan except Hokkaido. Small-scale aid would be provided to Nationalist Chinese forces if they still held substantial positions on the mainland, but no major offensive operations were contemplated in Asia.[87]

While performing their initial defensive tasks, the U.S. Air Force would launch a strategic air offensive against the USSR with atomic and conventional weapons. The primary target sets included atomic weapons production facilities, strategic air bases, key government facilities, major industrial areas, and petroleum industry sites. Operating from the continental United States, Okinawa, Alaska, England, and Cairo-Suez, ten bomber groups (439 aircraft) would begin their operations as soon as possible after D-Day. By D+1 month an additional two groups (sixty aircraft) would join the attack.[88]

The air offensive would materially reduce the Soviet Union's ability to sustain hostilities, but the Committee believed that air operations alone would not produce victory. Land and sea offensives would ultimately be necessary. The Allies would, therefore, mobilize their resources and after blunting the Russian attacks and reducing the Soviet military-industrial base mount a counteroffensive through the North German plain into Poland, the Baltic states and the northern USSR. The offensive would start from the Rhine or, if necessary, from British shores. The Committee rejected the old Pincher option of striking from Egypt through the Aegean Sea-Turkish Straits and Black Sea into the Ukraine. The southern attack would pose great strains on Allied logistics and lift capabilities, and only if the Soviets denied both Europe and the United Kingdom to the Allies, would the planners consider the southern option.[89]

The initial Dropshot concept did not provide details of either the air offensive or subsequent Allied offensive operations. It did, however, define Western Europe's defense as a major strategic requirement and attempted to describe necessary force levels and costs.

On March 7, 1949, the Air Force Chief of Staff, General Vandenberg, called for further detailed development of the plan's first phase.[90] General Bradley, the Army Chief of Staff, on March 18 also called for more detail, especially a discussion of the initial defense of the home territories of the European Allies. He also noted that the defense of Western Europe should be given equal priority with the defense of the Western Hemisphere.[91] Admiral Denfeld, the Chief of Naval Operations, expressed some reservations concerning Dropshot on April 4. He found the strategic concept for the defense of Western Europe too rigid and raised doubts about the ability to maintain Great Britain as a functional operating base area given enhanced Russian air power. He too joined his fellow service chiefs in calling for a more detailed examination of the problems of war in 1957.[92] On May 6, 1949, the Joint Chiefs directed the Committee to write an enlarged version of Dropshot. The JCS instructed the Committee to devise as a basic operational concept an approach for holding the United Kingdom and maximum areas in Western Europe, tasks that henceforth were deemed essential to the attainment of American political and strategic objectives.[93]

The *ad hoc* committee submitted a revised version of Dropshot to the JCS on December 19, 1949. The new version consisted of three volumes, and the Committee members noted that they had consulted with the Joint Strategic Survey Committee, the Joint Strategic Plans Committee, the Joint Logistics Plans Committee, the Joint Intelligence Committee, the Joint Military Transportation Committee, and the Budget Advisors to the Joint Chiefs of Staff.[94]

The Committee emphasized that Dropshot was a requirements study, a plan designed to determine the scale of effort and the cost of a global conflict. Between 1949 and 1957 the Committee estimated that if the United States wished to deter a war or fight on reasonably favorable terms it would have to spend twenty to twenty-two billion dollars a year of which two-and-a-half billion dollars annually would be spent on military aid to the nation's Allies.[95]

The rest of the first volume resembled the earlier Dropshot plan with a number of changes caused by new circumstances. Yugoslavia was no longer regarded as a loyal satellite, and its forces disappeared from the Soviet order of battle. The initial Allied defense line would be the Rhine-Alps-Piave position in Europe and in the Middle East the Allies would hold a line stretching from Crete to southeastern Turkey to the Tigris Valley and from there to the Persian Gulf. In Asia the Allies would defend Formosa, Okinawa, and Japan less Hokkaido.[96]

The volume also supplied details on force requirements for various phases of the war. On D-Day the U.S. Air Force required 3,529 aircraft, and the Allies needed 5,058 airplanes. By D+6 months the

American Air Force needed to expand to 4,270 and the Allies to 5,399 planes. By D+30 the U.S. Air Force would include 11,152 aircraft. The U.S. Navy would employ thirteen fleet, four light, and nine escort carriers, and the British would contribute two fleet and six light carriers. The U.S. Army would expand from about fifteen divisions during the opening phases of hostilities to seventy-four-and-two-thirds divisions by D+30 months. Along with Allied forces ground strength in all theaters by D+30 months would include 213-2/9 divisions and 701 antiaircraft battalions.[97] The Committee believed that its proposals for defense expenditures would make it feasible for the Western powers to reach these goals.

Dropshot's second volume dealt with Soviet capabilities and Allied reactions during the first months of hostilities. The volume differed only in matters of detail from the original report.

The Committee noted that all indications pointed to the fact that Moscow was not planning to launch a general war and sought to achieve its objectives by political means backed by military intimidation. War, however, could arise out of incidents between forces in direct contact or as a result of miscalculation.[98]

The planners believed that despite any problems the Soviet economy, transportation system, and morale were adequate to wage a major war.[99] Figures for the strength of the Russian armed forces were the same as those given in the initial study except for the Red Air Force. The Committee indicated that by D+180 days the Soviets would have 20,000 combat aircraft of which 9,400 would be jet propelled. This figure included 1,600 conventional and 400 jet powered long-range bombers. The Soviets would also have closed substantially the qualitative gap between their air arm and those of the United States and Great Britain.[100] The USSR would also possess by 1957 an atomic arsenal of about 250 bombs and a wide variety of antiaircraft rockets.[101]

In the absence of American military aid programs the USSR would be able to overrun much of Eurasia. By D+12 months the Red Army could conquer continental Europe, Turkey, the Middle East, South Korea, and Hokkaido. The Soviets would also launch atomic, biological, and sabotage attacks against the United States and Canada and seize for a short period the North Atlantic islands. Submarine attacks on merchant shipping would be mounted in both Atlantic and Pacific Ocean areas.[102]

Allied basic undertakings included: securing the Western
Hemisphere, north Atlantic Islands, United Kingdom, and as much of
Western Europe as possible. The Allies would conduct an air offensive
against the USSR and its satellites, destroy Soviet naval power, and
secure air and sea lines of communication. The United States in
particular would mobilize for total war and continue to provide military
aid to the Allies.[103]

In Europe the Allies would seek to hold the Rhine-Alps-Piave line.
The plan went on to provide a series of alternatives in case of failure to
accomplish each preceding course of action. In descending order of
acceptability the Allies would try to hold the Rhine-French-Italian
border, the Pyrenees-Apennines line, the Pyrenees line and Sicily, the
Pyrenees line, and the islands of Malta, Crete, and Cyprus. In the
Middle East the Committee concluded that the retention or early
reconquest of some of the region's oil resources was vital to the Allied
war effort. Thus, the Allies would attempt to establish a defensive
position stretching from Southeastern Turkey to the Tigris Valley and
from there to the Persian Gulf. Lacking sufficient forces, no effort
would be made to hold Norway, Sweden, and Denmark. In the Far East
the Allies would hold Japan less Hokkaido and Okinawa. Other regions
in southeast and southwest Asia were not immediately critical to the
West's war effort and would have to defend themselves.[104]

The third volume of Dropshot provided additional details
concerning Allied defensive strategy and also described with greater
precision American air defense efforts, the strategic air offensive, and
the counteroffensives of the war's later phases. The Committee also
provided additional data on conventional force deployments.

The Committee noted that no air defense system could be
completely effective. In order not to tie down forces needed elsewhere,
only vital installations would receive significant antiaircraft assets
supplemented by interceptor squadrons for area defense. Their basic
goal would be to force Russian long-range bombers to carry increasing
amounts of defensive equipment at the expense of bombs and to destroy
some of them in flight. The best defense of the Western Hemisphere
would be the strategic air offensive, which, if successful, would
substantially reduce Soviet capabilities to strike at North America and
to continue the war.[105]

In contrast to earlier plans for an air offensive which placed attacks
on Soviet industrial targets as the highest priority, Dropshot's first

priority was to neutralize the bases and facilities from which the Soviets would mount their long-range air attacks. Next the Air Force would concentrate on blunting Soviet land and sea offensives by attacking lines of communication, supply bases, troop concentrations directly supporting the Red Army's initial advances, and naval bases capable of supporting the USSR's submarine campaign. The third priority included attacks against the Soviet and satellite industrial economy.[106]

The atomic offensive would require nineteen American and seven British heavy- and medium-bomb groups with a total of 780 bombers, 144 reconnaissance planes and seventy-two weather reconnaissance aircraft. Seventy-five to 100 atomic bombs would be used against atomic assembly facilities, storage points, and heavy bomber airfields. The American and British air forces would deliver about 100 atomic bombs against tactical military and naval targets. Missions would be flown against satellite states and into areas overrun by the Red Army. These attacks would seek to avoid population centers. Conventional attacks would supplement the atomic offensive.[107]

For attacks on the Soviet military-industrial economy Dropshot designated three target-critical systems--the petroleum industry, the electric power system, and the iron and steel industry. The plan provided for the destruction of 75 to 85 percent of the petroleum industry, 60 to 70 percent of the electric power grid, and 75 to 85 percent of iron and steel producing facilities. The attacks would materially reduce the USSR's ability to sustain its war effort, and the by-products of the attacks, including loss of life, destruction of political and administrative centers, and disruption of communications, would have a significant impact on Russian morale.[108]

The planners calculated that to destroy the three systems and delay recovery for several years would require 180 atomic bombs in thirty days plus 12,620 tons of conventional bombs dropped by D+4 months. To destroy the electric grid and steel industry for one to one-and-a-half years would take 141 atomic weapons and 10,420 tons of conventional bombs. The destruction of the petroleum industry and electric power system for one year required 109 atomic bombs and 10,420 tons of conventional munitions.[109]

Attacks on similar target systems in satellite states called for seventy-three atomic bombs and 9,305 tons of conventional bombs to destroy the targets for several years. One to one-and-a-half years disruption required sixty-five A-bombs and 6,583 tons of conventional bombs, and to achieve damage for a year, the Allies had to drop sixty-four atomic bombs and 7,295 tons of high explosives.[110]

Thus, in the first thirty days of the war, SAC and RAF Bomber Command would drop a maximum of 453 atomic bombs on Soviet and satellite targets. In addition, by D+4 months the Allied air forces would drop 21,925 tons of high explosives. Bombers would operate from the continental United States, Alaska, Great Britain, Okinawa, and Cairo-Suez-Aden.[111] After the initial attacks, the Allies would continue to police the targets and if necessary reattack them.

The Navy required large forces to be ready on D-Day. Eight fleet and two light carriers with requisite escorts would operate in the Mediterranean. In the Barents-Norwegian Sea three heavy and three light carriers would be operational at the start of the war while four carriers would be located in the western Pacific.[112]

Of the seventy-six divisions assigned to hold the Rhine-Alps-Piave line the United States on D-Day would supply two infantry divisions. By D+30 days the U.S. contribution would grow to three infantry and one armored division, and by D+180 days American forces in Western Europe would consist of eight infantry and two armored divisions. In all the Allies would deploy sixty infantry and sixteen armored divisions supported by 4,107 aircraft.[113]

Anglo-American requirements in the Middle East by D+6 months included eleven Turkish, four British and three-and-a-third American divisions. Air support would consist of seven American and five British groups for a total of 650 fighters, ninety-six bombers and fifty-four reconnaissance planes.[114]

From the Persian Gulf to Japan the planners felt it unprofitable to deploy forces on the Asian mainland. Save for a British division and twenty-five fighters to protect Malaya, Soviet and Chinese moves would have to be met by local forces. Two American and ten Japanese divisions supported by 275 fighters would carry out the defense of Japan. The planners regarded these forces as adequate to hold the main islands and rejected the idea of defending Hokkaido.[115]

During the second phase of hostilities the Allies would continue the air offensive, maintain their ground positions, improving them where possible, and generate additional forces. The Allies would mobilize an additional ninety-seven divisions, fifty-five of which would be American. The U.S. Air Force by D+30 months would expand to ninety-four groups while the RAF and Commonwealth air arms manned thirty-three-and-a-third groups.[116]

If the USSR capitulated during Phase II, the Western powers would occupy Russia and the satellites in order to enforce their terms. Thirty-eight divisions and twenty-six air force groups would occupy selected cities stretching from Berlin to Vladivostok until civil governments resumed functioning.[117]

The Committee recognized that Moscow would probably not surrender as a result of the strategic bombing campaign. Consequently, the Allies would have to mount major land campaigns to destroy the Soviet and satellite armed forces.

Dropshot assumed that by D+24 months the Russian and satellite armies would contain about 650 divisions. The Soviets would have 125 divisions on the Rhine, thirty to forty along the North Sea coast, 125 in reserve, thirty-three in Anatolia, sixteen in the Iran-Iraq-Gulf region, and twenty in the Far East. Satellite divisions would be engaged primarily on occupation tasks in Germany, Austria, and the Balkans. The Soviets would also have over 5,000 combat aircraft. Russian forces would be suffering severe logistical problems because of Allied air attacks but would nonetheless retain a strong combat capability.[118]

The plan for the Allied counteroffensive proposed three operational alternatives: Plan North, Plan Pincher, and Plan South. All involved massive conventional campaigns designed to destroy enemy forces in Eastern Europe. Save for Moldavia and that portion of East Prussia annexed by the USSR in 1945, Allied forces would not launch major invasions of the Russian homeland. Rather, the planners hoped that the destruction of Soviet and satellite forces in Eastern Europe would force Moscow to capitulate.

Operation North involved the employment of 108 infantry, thirty-seven armored, and five airborne divisions supported by ninety air groups. About 114 divisions and seventy-three air groups would cross the Rhine north of Cologne and drive towards Berlin. Twelve divisions would land on the German North Sea coast, seize Bremen, Hamburg, and Lubeck, and deploy six divisions to liberate Denmark. Twenty-four divisions and fourteen air groups would later cross the Rhine south of Cologne and advance southeast towards Vienna.[119]

The main force would meanwhile link up with the amphibious force and take Berlin. Eighteen divisions would then advance and take the Baltic ports of Stettin, Danzig-Gdynia, and Konigsberg while forty-eight divisions moved east to capture Warsaw and Cracow. The Vienna attack force would advance at this time and link up with units advancing on Cracow.[120]

From Cracow, twenty-four divisions supported by fourteen air groups would advance southeast with the objective of reaching the Black Sea and seizing the mouth of the Danube. The force would advance down the corridor formed by the Carpathian Mountains and the Dniester River. The final advance would isolate Soviet and satellite forces in Western and Central Europe from the USSR.[121]

Operation Pincer called for an eastward advance from the Rhine coupled with an assault through the Aegean, the Turkish Straits and Black Sea. Pincer required 117 infantry, thirty armored, and nine airborne divisions, and 110-1/3 combat air groups.[122]

The southern arm of the Pincer would strike first. Allied air power would neutralize Soviet naval and air capabilities in southern Turkey and Greece. A force of twenty-seven divisions supported by twenty-five-and-a-third air groups and twelve escort carriers would seize Athens-Piraeus, Macedonia, Thrace, and the Turkish Straits. After consolidating their gains, Allied forces, reinforced to a strength of fifty divisions, would send thirty of them backed by twenty-six air groups and twelve carriers in a two-pronged thrust aimed at Bucharest. One force would advance overland through Bulgaria while the other would execute an amphibious landing on the Rumanian Black Sea coast. Finally, the Allies would mount a major drive through Rumania towards Cracow while other units established flanking positions on the Dniester and the Transylvanian Alps.[123]

Two to three months after launching the southern campaign, the Allies would strike in the north. Seventy-eight divisions and fifty-two air groups would strike for Berlin; twelve divisions would land on the North Sea Coast and liberate Denmark, and twenty-four divisions and fourteen groups would drive towards Vienna. From Berlin fourteen divisions would advance on the Baltic ports; twenty would move on Warsaw; and twenty would seize Cracow and link-up with forces advancing to and beyond Vienna. Forces in Cracow would then push south and join with the southern arm of the Pincer.[124]

Operation South called for a main offensive through the Aegean, the Turkish Straits, the Rumanian Black Sea coast, and from there into Eastern Germany in conjunction with holding operations on the Rhine.

Cracow, Warsaw, the Baltic ports, Berlin, Dresden, and the line of the Elbe River would be seized from the south in a series of operations requiring 182 divisions and 124-1/3 combat air groups.[125]

The Committee recommended Operation South only as a last resort. They noted that the operation required more forces than the other alternatives, and the time required to generate them would cause excessive delay in launching the offensive. Furthermore, the Committee pointed out that because of the long sea and land lines of communication the logistic feasibility of Operation South was open to serious question.[126]

Of the remaining two alternatives the planners recommended Operation Pincer. Operation North also involved long communications lines at the end of the campaign and faced the bulk of the Red Army. Operation Pincer would be easier to sustain logistically, would force the Soviets to split their forces, and would prevent Soviet forces from escaping back to Russia via the Balkans.[127]

Although designed as a requirements plan, Dropshot was a massive study involving the participation of almost all of the JCS Committees. The plan accepted the new American commitment to defend Western Europe. European defense in fact became a leading priority replacing concern for Cairo-Suez and Middle East oil regions. These areas continued to be important, but their value was viewed as less critical than Europe west of the Rhine.

Massive conventional forces were indispensable for fighting the USSR and even the atomic offensive would not eliminate the need for regular army, navy and air force units. Atomic warfare also evolved. No longer did the plan call for attacks on Soviet cities in the hope of doing collateral damage to industrial systems. Dropshot emphasized counterforce targets--Soviet atomic building and delivery systems and tactical targets as its leading priorities.

The initial American contribution to the Allied war effort would be the strategic air offensive and military aid. The Allies would supply the bulk of the ground forces with the American contribution growing in the later phases. As with earlier plans, any war against the USSR would be global, total and unlimited.

Dropshot was never formally approved by the JCS, but it remained "on the books" until 1951. It provided a good outline of what the military thought was necessary to resist Soviet power and indicated the costs of programs to create the required forces. Ultimately overtaken by events, Dropshot did summarize American strategy in the context of a permanent partnership designed to protect Western Europe.

NOTES

1. Kenneth W. Condit. *The History of the Joint Chiefs of Staff, The Joint Chiefs of Staff and National Policy, Volume II 194–1949*. Wilmington, Delaware: Michael Glazier, Inc., p. 561, and Steven L. Rearden. *History of the Office of the Secretary of Defense: The Formative Years 1947–1950*. Washington, D.C.: Historical Office of the Secretary of Defense 1984, pp. 335–351. *See also*: *Second Report of the Secretary of Defense 1949*. Washington, D.C.: GPO, 1949, pp. 134–135 and 196.
2. JIC 380/2, 16 February 1948.
3. JIC 429/1, 25 August 1948.
4. JIC (48) 70 (0) Final, 21 September 1948.
5. JIC 435/12, 30 November 1948.
6. *Ibid.*
7. *Ibid.*
8. *Ibid.*
9. JCS 1837, 20 February 1948, and JCS 893/5, 24 May 1948.
10. JIG 286/1, 22 March 1948.
11. JIG 286/2, 17 September 1948.
12. SANACC 398/1, 6 May 1948, and SANACC 398/4, 25 May 1948.
13. *Ibid.*
14. JCS 1745/5, 8 December 1947.
15. JIC Policy Memo no. 2, 29 March 1947.
16. JCS 1828/1, 17 April 1948.
17. JCS 1745/15, 27 July 1948, and JCS 1823/14, 27 May 1949.
18. David Rosenberg. "American Atomic Strategy and the Hydrogen Bomb Decision," *Journal of American History*, Vol. 66, no. 1, June, 1979, p. 71.
19. *Ibid.*, pp. 72–73.
20. *Ibid.*, p. 72.
21. *Ibid.*, p. 73.

22. *Ibid.*, p. 75.
23. JCS 2084, 16 November 1949.
24. JCS 2084/2, 31 January 1950.
25. JCS 1868/1, 17 April 1948.
26. JSPC 876, 21 April 1948.
27. *Ibid.*
28. JSPC 877/9, 21 July 1948.
29. JCS 1868/22, 1 October 1948.
30. JCS 1868/63, 8 March 1949.
31. Metric, 26 April 1949, and Metric, 6 July 1949.
32. JCS 1868/99, 16 August 1949.
33. JCS 1868/86, 28 May 1949, and JCS 1868/16, 5 August 1949.
34. Condit, *op. cit.*, pp. 295–296.
35. JCS 1844/37, 27 April 1949.
36. *Ibid.*
37. *Ibid.*
38. *Ibid.*
39. *Ibid.*, and JCS Meetings of 13 May 1949 and 17 June 1949.
40. JCS 1844/46, Joint Outline Emergency War Plan Offtackle, 8 November 1949. (The name of the plan was later changed to *Shakedown* and still later to *Crosspiece*.)
41. *Ibid.*
42. *Ibid.*, IV National War Objectives.
43. *Ibid.*
44. *Ibid.*
45. *Ibid.*, III Assumptions.
46. *Ibid.*
47. *Ibid.*
48. *Ibid.*, VIII Estimate.
49. *Ibid.*
50. *Ibid.*, First Phase Tasks-Defensive (D to D+3 Months).
51. *Ibid.*
52. *Ibid.*
53. *Ibid.*
54. *Ibid.*
55. *Ibid.*
56. *Ibid.*
57. *Ibid.*
58. *Ibid.*

59. *Ibid.*
60. *Ibid.*, and Rosenberg, *op. cit.*, 71.
61. *Ibid.*
62. *Ibid.*, and unpublished notes from research by Professor David Rosenberg.
63. *Ibid.*, Second Phase (D+3 to D+12 Months).
64. *Ibid.*
65. *Ibid.*
66. *Ibid.*, Third Phase (D+12 to D+24).
67. *Ibid.*
68. *Ibid.*, Fourth Phase (D+14 to the End of the War).
69. JCS 1844/47, Report by the Joint Logistics Plans Committee to the Joint Chiefs of Staff on Logistic Implications of "Offtackle," 15 November 1949.
70. JCS 1844/53, 26 January 1950.
71. JCS 1844/55, 18 February 1950.
72. JCS 1920, 16 August 1948.
73. JCS 1920/1, Long-Range Plans for War with the USSR—Development of a Joint Outline Plan for use in the Event of War in 1957 (Short Title: "Dropshot"), 31 January 1949.
74. *Ibid.*, Special Assumptions.
75. *Ibid.*
76. *Ibid.*, Relative Combat power.
77. *Ibid.*, USSR and Satellites (2) Attitude and Morale.
78. *Ibid.*, Logistics.
79. *Ibid.*, Probable Soviet Strategic Objectives.
80. *Ibid.*, Allied Courses of Action.
81. *Ibid.*
82. *Ibid.*
83. *Ibid.*
84. *Ibid.*
85. *Ibid.*
86. *Ibid.*
87. *Ibid.*
88. *Ibid.*
89. *Ibid.*
90. JCS 1920/2, 7 March 1949.
91. JCS 1920/3, 18 March 1949.
92. JCS 1920/4, 4 April 1949.

93. JCS 1920/1, 6 May 1949.
94. JCS 1920/5, Long-Range Plans for War with the USSR—Development of a Joint Outline Plan for use in the Event of War in 1957 (Short Title: "Dropshot"), Volume I, 19 December 1949.
95. *Ibid.*
96. *Ibid.*
97. *Ibid.*
98. *Ibid.*, Volume II, Political Factors.
99. *Ibid.*, Soviet and Satellite Armed Forces.
100. *Ibid.*
101. *Ibid.*
102. *Ibid.*, Soviet Capabilities.
103. *Ibid.*, Allied Courses of Action.
104. *Ibid.*
105. *Ibid.*, Volume III, Development of Tasks and Force Requirements (Phase I).
106. *Ibid.*
107. *Ibid.*
108. *Ibid.*
109. *Ibid.*, Appendix "C" to Enclosure "K."
110. *Ibid.*
111. *Ibid.*, Development of Tasks . . . and Appendix "C" to Enclosure "K."
112. *Ibid.*, Development of Tasks.
113. *Ibid.*
114. *Ibid.*
115. *Ibid.*
116. *Ibid.*
117. *Ibid.*
118. *Ibid.*, Phase III and Volume II, Soviet Capabilities.
119. *Ibid.*, Vol. III, Phase III.
120. *Ibid.*
121. *Ibid.*
122. *Ibid.*
123. *Ibid.*
124. *Ibid.*
125. *Ibid.*
126. *Ibid.*

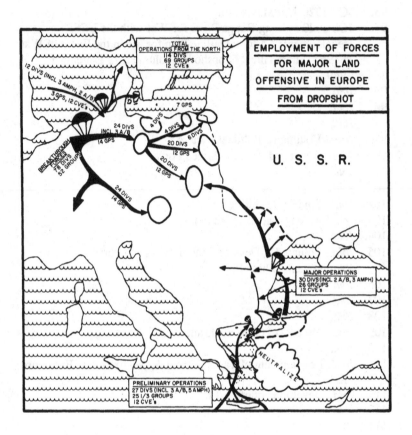

TOTAL
OPERATIONS FROM THE NORTH
114 DIVS
69 GROUPS
12 CVE's

EMPLOYMENT OF FORCES

FOR MAJOR LAND

OFFENSIVE IN EUROPE

FROM DROPSHOT

12 DIVS (INCL 3 AMPH, 2 A/B)

3 GPS, 12 CVE's

7 GPS

24 DIVS
INCL 3 A/B
14 GPS

4 DIVS

4 DIVS   6 DIVS

BREAKTHROUGH
FORCES
78 DIVS
52 GROUPS

20 DIVS
12 GPS

20 DIVS
12 GPS

U. S. S. R.

24 DIVS
14 GPS

MAJOR OPERATIONS
30 DIVS (INCL 2 A/B, 3 AMPH)
26 GROUPS
12 CVE's

NEUTRALIZE

PRELIMINARY OPERATIONS
27 DIVS (INCL 3 A/B, 5 AMPH)
25 1/3 GROUPS
12 CVE's

CHAPTER VI

The Continuing Dilemma

By 1950 American military planners, following the evolution of national policy, regarded the defense of Western Europe as an essential strategic task. Europe's defense was vital to the security of the United States. Thus, U.S. planning had come full circle. Before America's entry into World War II, Army and Navy strategists identified the Atlantic-European area as the decisive theater. During the first years of the Cold War it did not seem feasible to defend Europe against a Soviet attack. The JCS, therefore, substituted the concept of waging offensive operations in Western Eurasia, but what they really meant was that Anglo-American forces would launch air and, ultimately, amphibious campaigns from the British Isles and the Middle East. Administration policies, however, led the JCS to restore the defense of Europe to its dominant position in American strategy.

Plan Offtackle and Plan Dropshot were the first responses to the new strategic orientation. Offtackle, using existing forces, had to relinquish Europe but intended to reinvade the Continent once the Allies generated sufficient new forces. Dropshot sought to establish requirements for an effective forward defense and for a decisive counterblow. Dropshot presumed greatly increased levels of future defense spending, but planners dealing with immediate strategic problems remained caught in a familiar quandary—the need to respond to expanded national commitments with insufficient conventional forces. The standard response was to call for larger budgets and to rely on atomic weapons, but the government often balked at increasing defense expenditures, and the development of a Soviet atomic arsenal and modernization of the Red Air Force and air defense network raised a number of questions about the Strategic Air Command's ability to inflict heavy damage upon the USSR and about the prospect of the atomic air offensive playing the decisive role in a future Soviet-American global conflict.

American estimates of the military balance continued to emphasize Russian power. An appraisal of the current international situation of September 3, 1949, noted that the USSR would not deliberately resort to war, but if a conflict did erupt, the Red Army would overrun Western Europe in two months, take Iran, Iraq, and Syria in three months and be in Cairo in six months.[1] Nevertheless, on October 25, 1949, the JCS noted that NATO strategy would attempt to hold the Soviets on the Rhine or further east if possible and would also defend a Scandinavian bastion.[2]

A Joint Advanced Study Committee in mid-February, 1950, indicated that the United States would continue to rely heavily on atomic weapons to achieve its strategic objectives. The Committee asserted that because of the destructive power of atomic weapons the opening phases of a war would be both violent and critical. To hold the line of the Rhine the Allies would employ small-yield tactical atomic weapons. At the same time the U.S. Air Force would seek to neutralize the Russian strategic air offensive by hitting atomic weapons production plants, storage sites, and bases for long-range bombers. The Allies would also destroy Middle Eastern oil refineries with atomic and conventional munitions to deny their use to the Russians.[3]

In late February an Intelligence Committee report emphasized the magnitude of Russia's armed might. The Red Army could on D-Day place 175 line, thirty-five artillery and fifty-six satellite divisions in the field. The Soviet Air Force had 20,100 aircraft, including 1,725 long-range bombers, and the Navy possessed 400 submarines and 3,225 planes.[4] American forces available to execute Plan Offtackle included one division and three regiments in Europe and five divisions and five regiments in the United States. The Navy could deploy two light and four fleet carriers in the Atlantic, and the Air Force had available fourteen bomber and six-and-two-thirds fighter groups.[5] The full NATO alliance deployed ten divisions in West Germany, and planners estimated that eighteen were required to conduct an effective fighting retreat to the Rhine. Field Marshal Montgomery reported in mid-June that the Allies had little chance, given current force ratios, of offering effective resistance to the Red Army.[6]

Increased defense spending was one solution to the imbalance of forces, and in early 1950 the National Security Council ordered a study of America's diplomatic and military position. The Joint Strategic Survey Committee represented the JCS on an interdepartmental study group which at the end of March presented a report, NSC68, that called for the military budget to grow from $13 billion to $35 or even $50 billion.[7]

Military spending did increase from $14.258 billion in FY1950 to $53.208 billion in FY1951 and to $65.992 billion in FY1952. The JCS began to contemplate a twenty-one division Army, a 143 wing Air Force and a 402 combat ship Navy by FY1952.[8] President Truman, however, never sent NSC68 to Congress, nor did he officially approve it until September 1950, and it was the Korean War rather than NSC68's call for rearmament in peacetime which produced major increases in the defense budget. In the absence of the Korean War military growth might well have been far more limited, and the end of the war promised a return to fiscal austerity. Moreover, much of the additional spending was devoted to the Asian conflict and did not dramatically increase forces available for campaigns in Europe and the Middle East.

Since the prospects for peacetime rearmament were ambiguous, American strategists had to continue to rely heavily upon atomic weapons, especially in the early stages of a global conflict. Despite a growing arsenal and, under General Curtis Le May's dynamic leadership, a larger, more efficient Strategic Air Command, doubts expressed in 1945 and 1946 concerning delivery capabilities and the strategic impact of atomic weapons persisted.

On December 31, 1949, SAC had 521 aircraft able to deliver atomic bombs and a total inventory of 837 planes. Two years later the numbers had increased to 658 bombers and 1,165 total inventory.[9] A Weapons System Evaluation Group report of early February, 1950, indicated, however, that an atomic offensive would suffer heavy losses. The report noted that under the most favorable assumptions seventy to eighty-five percent of the aircraft sortied would successfully attack their targets. In night raids SAC would suffer about thirty-five percent casualties, and in daylight raids the loss rate would climb to fifty percent. WSEG concluded that anticipated losses and logistical shortages would make it impossible to execute the air offensive called for in Offtackle. SAC could deliver its 292 atomic bombs but would have to abandon the follow-up conventional attacks.[10]

An Intelligence Committee report of August 22, 1950, stated that the USSR possessed twenty atomic bombs and would have 165 by 1953. American superiority in atomic weapons was, therefore, a dwindling asset. The report went on to note that Moscow did not intend to unleash a general war, but if Russia ever did decide to strike, the most favorable period would be between 1951 and 1953.[11] In February, 1951, the Joint Intelligence Committee informed the JCS that the Red Army with 175 divisions supported by 20,000 aircraft was the best in the world. In 1952 the Russians would have 120 atomic bombs, and Soviet strength in relation to the West would be at its peak in 1952.[12] The USSR could support logistically multiple offensives, neutralize much of America's strength with its atomic weapons and because of prewar stockpiling carry on with its campaigns during the first year of hostilities despite American air attacks.

Thus, American atomic capabilities had important but not decisive utility and were not the sole solution to the West's strategic problems. U.S. military planners had, however, little choice but to rely upon atomic weapons to counter Soviet conventional superiority even though the expanding Russian atomic arsenal posed a growing threat. President Truman's decision at the end of January, 1950, to develop the hydrogen bomb offered no immediate solution to the military's basic problem of devising an effective strategy to defend Europe and the Middle East with severely limited forces.

The JCS continued to regard Asia as a secondary defensive theater. On January 13, 1950, the JIC noted that for the Soviets, Asia was also a secondary area. In case of war the Russians would attempt to suck Allied forces into Asia primarily to prevent their deployment in Europe.[13]

The Korean War did not substantially change the JCS point of view. On June 29 the Chiefs informed General MacArthur that if Soviet forces entered Korea his troops should defend themselves but take no action to aggravate the situation. The U.S. would also defend Formosa but prevent Nationalist forces from launching major attacks against the mainland.[14] The following day, the JCS noted that Soviet intentions in Korea were to discredit American credibility, discourage

NATO members, test America's will to sustain Allies, and gain strategically useful territory. Korea was not, however, the start of a general war.[15]

If Korea were the first campaign in a global war, the Joint Chiefs stated that American strategy called for a retreat from the peninsula and the execution of Plan Offtackle.[16] On July 12, the Joint Strategic Plans Group reported that although the Korean War prevented the deployment of additional American forces to Europe the United States still remained committed to Offtackle. Furthermore, the U.S. hoped ultimately to become strong enough to defend the Rhine, and the response in Asia should serve as an example to the European Allies of the American willingness to honor national obligations.[17]

The government wished to keep the Korean War limited, and on October 3, 1950, the JCS instructed General MacArthur that if his forces were north of the 38th parallel and encountered Russian or Chinese troops his command was to assume defensive positions, and he was to report any encounters to Washington. South of the parallel MacArthur's forces could continue to advance if they encountered Chinese troops, but if they clashed with Russian units, they were to halt and assume a defensive posture.[18] In late December the Joint Chiefs concluded that in case of open war with China the U.S. must not lose sight of the fact that the USSR was the main enemy. The United States would, therefore, try to stabilize the situation in Korea with the forces presently there and initiate a blockade and an air offensive against the Chinese mainland. The U.S. would also assist the French in Indo-China, build up Nationalist forces for a reinvasion, assist guerrilla activities on the mainland, and strengthen Japanese forces.[19] The JCS would not, however, fight a major war in Asia to the detriment of Europe's defense.

The JCS reaffirmed the commitment to the strategy of Offtackle during the next two years. In January, 1951, the JCS called for stabilizing the line in Korea but except for replacements no additional forces would deploy to Asia. If the line could not be held with current forces, the Allies would evacuate the peninsula.[20] In February the Joint Strategic Plans Committee reasserted that American strategy called for the stabilization or evacuation of Korea and a build-up of Japanese forces.[21] In June the JCS stated that if the Soviets initiated hostilities in the Far East as the first move in a global war, the U.S. would adhere to the Offtackle strategy by abandoning Korea, defending Japan, and launching a strategic air offensive.[22] In April, 1952, the JCS decided that in case of limited US-USSR hostilities in the Far East the Americans would retreat from Korea and defend Japan, Okinawa, Formosa, and the Philippines. The United States would also conduct an atomic and conventional offensive in the Far East and if the war became general revert to the defense in Asia.[23] In July the Far East Command argued that the Soviet Union was probably planning to stand on the defense in the west in order to launch a one-front war in Asia. The U.S. should, therefore, reinforce its position in the region to a point where it could fight a major Asian war.[24] The JCS rejected this proposal as well as additional suggestions from the Far East Command for a greater emphasis on operations in Asia.[25]

For American strategists, Europe remained the region of primary concern, but the conventional defense of Western Europe continued to suffer from substantial shortages of men and material. The JCS on a number of occasions called for the rearmament of West Germany in order to expand the Allied order of battle.[26] The JCS also agreed to increase the number of American troops in Western Europe, bringing the U.S. contingent to four infantry and one-and-a-half armored divisions and eight or ten tactical air groups. Four divisions would deploy to Europe in 1951.[27]

Despite efforts to improve the American and Allied force posture, serious shortages remained. In August, 1950, the JCS noted that plans for defense forward of the Rhine were militarily unsound because NATO lacked 8,000 tanks, 9,200 half-tracks and about 3,200 artillery pieces. In September the JCS noted that NATO needed an additional 8,636 aircraft for tactical and defensive missions.[28] Thus, political ambitions and the military force necessary to achieve them remained uncoordinated and plans continued to be inconsistent with real orders of battle.

While dealing with current problems, the JCS on January 26, 1950, ordered the Joint Strategic Plans Committee to prepare a mid-range war plan. On November 29, 1950, the Joint Chiefs approved Plan Reaper which presumed a conflict beginning on July 1, 1954.[29] Reaper was the first war plan written against the background of NSC68 and the increased defense spending that emerged during the Korean conflict.

Reaper was closer to Dropshot than to Offtackle in its overall strategic view. The planners assumed that as a result of continued rearmament a forward defense of Western Europe and the Middle East would be feasible and was not merely an empty gesture designed to reassure the Allies. Unlike earlier plans, Reaper also presumed the active participation of a large West German army and the existence of a substantial Soviet atomic arsenal.[30]

Although the plan provided no description of how the war would start or why Moscow would decide to launch World War III, Reaper assumed that the conflict would open with several Soviet offensives designed to overrun Western Europe and the Middle East. In the Far East the Russians would launch attacks designed to pin down Allied forces in the region. Moscow would also launch worldwide sabotage and subversive campaigns and unleash its submarines against Allied shipping. The USSR would also use its 235 atomic bombs to cripple the industrial base of England and the United States. Thirty to forty A-bombs would be launched against the United Kingdom; the rest would fall upon the United States. The Russian atomic attack would inflict great though unspecified damage on both nations.[31]

The Allied strategy as in other plans involved protecting the industrial base of the Western Hemisphere, guarding vital sea and air lines of communication, deploying to advanced bases including England, Iceland, the Azores, North Africa, and Cairo-Suez-Aden, and launching an air offensive against the Soviet Union.[32] The primary target systems for U.S. atomic weapons were: the destruction of the Soviet air atomic offensive capability, supporting the defense of Western Europe by attacking targets vital to the support of the initial Russian offensive, and attacking the Soviet industrial base with emphasis on petroleum and transportation facilities.[33] The Strategic Air Command would launch its first atomic attacks on approximately D+6 days. SAC intended to hit the Moscow-Gorky area with twenty bombs and Leningrad and vicinity with twelve weapons. SAC would also drop fifty-two bombs in the Volga and Danets Basin industrial regions, and fifteen bombs would be used to hit the Caucasus area. In the Far East SAC would deliver fifteen bombs against Vladivostok and Irkutsk.[34]

The Allies would also establish a defensive line. In Europe NATO forces including ten West German divisions would hold the Red Army as far to the east as possible. Demolitions, guerrilla warfare, and a maximum air effort with conventional and atomic weapons would at least enable the Allies to establish a defensive position along the line Trondheim-Oslo-Zeeland-Rhine-Alps-Piave River.[35] Thus, in contrast to all previous plans the Allies would seek to hold a Scandinavian bastion.

In the Middle East the Allies would protect the Cairo-Suez-Aden base region and the region's oil resources by holding a line from southeastern Turkey to the Iranian mountain passes and from there to the Persian Gulf.[36] Reaper was the first plan to propose holding a line north of the Gulf. The planners presumed that Soviet and satellite armies would overrun Greece and Yugoslavia but that Crete and Cyprus could be held, thereby linking the European and Middle Eastern fronts.

In the Far East the Allies would establish a line from the Kra Isthmus along the China coast to the Bering Strait thus protecting Japan and Okinawa. About ten Japanese divisions would assist American forces.[37] The planners intended to hold Japan and Malaya but curiously ignored the defense of both Korea and Formosa.

Allied defensive operations would occupy the first three months of hostilities. The second phase from D+3 to D+9 months would witness continued operations to attain air superiority and to secure critical lines of communication. The United States and the Allies would also continue to mobilize their resources, a process begun even before D-Day.[38]

Phase III, lasting from D+9 to D+24 months, would involve the completion of mobilization and definitive attainment of air superiority. The LOCs would be fully secure, and the Allies would also undertake limited counteroffensives.[39] Reaper did not, however, describe the nature, size or operational goals of these attacks.

The fourth phase, D+24 to D+48 months, would mark the inauguration of major Allied offensive operations. Allied forces would be substantial. Mid-way through the fourth phase, the American Army would consist of eighty divisions, seven regiments, and 236 AA battalions. The Air Force would contain 211-1/3 wings, and the Navy would operate sixteen fleet and thirty-six escort carriers.[40] The British would field thirty-one divisions, forty-nine-and-a-third air wings, and seven carriers, and the French forces would include thirty-three divisions, forty-three air wings, and two light carriers.[41]

Unlike Dropshot the planners were vague about the Allied offensives. There would be major operations through the North German plain, but Reaper supplied no information concerning the size or objectives of this offensive. The Allies would also mount attacks in the Mediterranean and Aegean to support the thrusts in the north, but again Reaper said nothing about the specific strategic objectives of the southern attacks. Finally, the Allies would prepare to conduct alternative strikes into Italy and/or Yugoslavia and into Scandinavia. Reaper was, however, silent on the issues of size, timing, or objectives.[42]

The Reaper Plan assumed that the American build-up called for in NSC68 would be accepted by the President and Congress. It also assumed that the NATO allies would increase their levels of defense spending and that West German and Japanese rearmament would be both politically and economically feasible.

In contrast to earlier war plans Reaper contained no detailed information on Soviet or satellite strength, and the description of Moscow's strategy was vague and general. Reaper did, however, go into great detail on military requirements for various phases of the war in order to provide guidance for mobilization planning and budget requests.

Despite JCS approval Reaper caused much dispute. In October, 1951, General Vandenberg became convinced that the plan no longer offered adequate strategic guidance, and the Chiefs of Staff directed the JSPC to decide if Reaper needed revisions or a complete rewriting. The Committee and ultimately the Joint Chiefs became mired in arguments centering around the importance and priority of essential strategic tasks. The Army and Navy argued that essential tasks including the protection of the Western Hemisphere, the security of the NATO area, the atomic offensive, and the defense of other critical areas were equally important, and any war plan had to meet each threat. The Air Force insisted that threats and responses had to be given orders of priority and that the air offensive against the USSR's war-sustaining capability was the U.S. military's most critical task.[43]

General Collins, Army Chief of Staff, offered a compromise by listing broad war-planning priorities with the understanding that they were not successively exclusive. The CNO, Admiral William M. Fechteler, however, continued to insist that all basic missions were equally important, and the failure to perform any one of them might lead to the collapse of the entire war effort. The JCS then asked the JSPC whether the opinions of General Collins and Admiral Fechteler could provide adequate guidance for war planning. In September, 1952, the Committee returned a negative response. The JCS, finally, decided to retain Reaper subject to changing the force projections to respond to new budget requests and intelligence data.[44]

The Chiefs of Staff also decided to revise the entire planning cycle. In January, 1950, the JCS called for the annual preparation of an emergency, a mid-range, and a long-range plan. The coordinated program, after many delays, was introduced in July, 1952. It called for a Joint Strategic Capabilities Plan (JSCAP) to deal with the coming fiscal year, a Joint Strategic Objectives Plan (JSOP) to provide requirements and strategic guidance for a war beginning three years after the plan's completion, and a Joint Long-range Strategic Estimate (JLRSE) to apply to a five-year period beginning five years after the plan's issuance.[45] The new planning system did not resolve inter-service disputes but did establish a consistent coordinated planning cycle which is still largely in use today.

Long- and mid-range plans, no matter how optimistic, could not, of course, solve the immediate strategic dilemma of meeting expanded political commitments with insufficient forces. During the years following Reaper, shortages in conventional forces became a constant litany.

On July 30, 1951, the JCS noted that serious force gaps remained in NATO, thereby rendering the defense of West Germany unfeasible.[46] In August the JSPC reported that NATO had not yet met the requirements necessary to execute current defense plans.[47] On September 12, the JCS reported that on D-Day NATO required forty-six divisions but could muster only thirty-one. NATO required ninety divisions by D+90 but could mobilize only sixty-seven.[48] Other reports indicated that the U.S. and the Allies lacked sufficient reserves of aircraft fuel after D+45 days and that for eighteen months after D-Day American divisions would have to fight without full equipment inventories.[49] At the end of October, the JCS approved a Joint Intelligence Committee report stating that forces for a forward strategy were inadequate.[50]

The JCS recognized the virtual impossibility of defending the Rhine and in early November, in guidance for emergency war planning, instructed planners to prepare for a fighting retreat to the Rhine and from there to the Pyrenees. The Allies would also undertake a fighting withdrawal down the Italian boot.[51] In the Middle East the Allies would hold Cairo-Suez and other areas if possible.[52] No attempt would be made to hold the Iranian mountain passes or even the south shore of the Persian Gulf, for as the JLPC pointed out more than a year later,

American armed forces were despite increased defense spending still critically short of aircraft, specialist personnel, petroleum stocks, and ammunition.[53]

On December 28, 1951, the JIC noted that the Allies would face very powerful Soviet forces including 175 line divisions, 300 submarines, and over 20,000 aircraft. The Russians could support logistically attacks against Western Europe, Scandinavia, Italy, the Balkans, Turkey, and the near East. The Intelligence Committee also credited the Soviet Air Force with the ability to hit the United States and Canada.[54] The following day, the JSPC again noted that in case of a Soviet attack and despite an informal agreement between Montgomery and Tito to deploy a Yugoslav corps in defense of the Ljubljana Gap, the Allied final defense line would rest on the Pyrenees and on a bridgehead in southern Italy.[55]

The JCS response to their strategic dilemma was to advocate greater defense efforts by the NATO allies, to call vigorously for West German rearmament and to supply the Supreme Allied Commander Europe with tactical atomic weapons. By January 1952, SACEUR had 200 bombs allotted to his command.[56]

Results, however, were mixed. The Allies did promise to expand their forces although performance often fell short of Alliance goals. West Germany, after years of often bitter debate, did create a powerful army, and the tactical nuclear stockpile ultimately expanded from 200 to many thousands of weapons.

Despite these efforts the USSR appeared to maintain a decisive lead in conventional forces. In June, 1953, the Joint Intelligence Committee noted that the USSR kept thirty divisions in Eastern Europe and could quickly reinforce them with eighty-two divisions from European Russia. Moreover, the Soviets could quickly mobilize another twenty divisions and expand the Red Army to 320 divisions by D+30. The Committee estimated that the Russians by 1954 could support logistically 175 to 200 divisions for an attack against Western Europe.[57]

Later in the month the JIC revised its estimates. The Soviets currently could supply 138 divisions in a European campaign supported by 12,000 tactical aircraft. The Russians were, however, increasing their war-reserve stockpiles and improving their lines of communication. When completed, the Red Army would be able to sustain as many as 200 Soviet and satellite divisions.[58]

The JCS also felt that the U.S. was becoming vulnerable to atomic attack and noted that by 1957 the USSR would have over 500 A-bombs and the means to deliver them to the North American continent. The Joint Chiefs nevertheless believed that despite their inferiority in conventional forces and eroding security from atomic attack, war was unlikely. The American atomic deterrent was still powerful and the USSR was in the final analysis a cautious power that preferred to expand by means short of war. Moscow was, therefore, unlikely to launch a war as a calculated policy. A Russian attack, concluded the JCS in October, 1953, was not imminent.[59]

The political optimism, that the Soviet Union would not launch a general war, and the military pessimism, that if Moscow struck America and the Allies would be hard pressed, that characterized JCS strategic views in the immediate post-war years continued into the early 1950s. The Red Army in terms of forces in being shrank but obtained new equipment and remained in the American view the world's best army. The Soviet Air Force moved rapidly into the jet age, and the navy expanded its oceangoing submarine fleet. Moscow also began to develop an atomic capability. Nevertheless, the JCS felt that domestic imperatives, the need to repair war damage and modernize the industrial economy, plus the threat of atomic retaliation would convince Stalin to continue to try to avoid war.

During the early 1950s new alliances provided advanced bases for SAC and held the promise of someday being able to create conventional forces sufficient to match those of Russia. NATO did not, however, in its early years fulfill its potential, and JCS planners had to accept the brutal fact that in the short term the Alliance could not feasibly create an effective defense for most of Western Europe. Mobilization and future counteroffensives were the only alternatives.

Planning did shift its focus from the Middle East to Europe. Mid-term and long-range plans, assuming that there was adequate funding, called for the execution of a forward defense of the NATO area, but such a defense was not feasible at the beginning of 1953. Nevertheless, the implicit understanding that Soviet policy was guided not only by the military balance but also by broader economic and political considerations led the JCS to conclude that, adverse indicators to the contrary, war in the near future was not probable.

NOTES

1. JCS 1924/6, 3 September 1949.
2. JCS 1868/136, 25 October 1949.
3. Joint Advanced Study Committee 508, Pattern of War in the Atomic Age, 15 February 1950.
4. JCS 2073/7, 27 February 1950.
5. JSPC 876/124, 18 May 1950.
6. Walter S. Poole. *The History of the Joint Chiefs of Staff. The Joint Chiefs of Staff and National Policy, Volume IV, 1950–1952*. Wilmington, Delaware: Michael Glazier Inc., 1980, p.185.
7. *Ibid*, pp. 4–8.
8. *Ibid.*, pp. 107 and 137.
9. *Ibid.*, p. 168. JCS 1823/28, 14 July 1950 gives different figures: 381 Air Force bombers and sixteen naval aircraft.
10. Poole, *Op. cit.*, pp. 163–164.
11. JIC530/3, 22 August 1950.
12. JIC435/52, 7 February 1951.
13. JIC500, 13 January 1950.
14. JCS 1776/6, 29 June 1950.
15. JCS 1924/10, 30 June 1950.
16. JCS 1776/13, 2 July 1950.
17. JCS 2073/41, 12 July 1950.
18. JCS 1776/122, 30 October 1950.
19. JCS 2118/4, 27 December 1950.
20. JCS 2118/4, 12 January 1951.
21. JSPC 958/15, 16 February 1957.
22. JCS 1924/59, 26 June 1951.
23. JCS 1924/65, 11 April 1952.
24. JCS 1924/67, 24 July 1952.
25. JCS 1924/68, 7 August 1952; JCS 1924/69, 25 November 1952; and JCS 1924/70, 19 February 1953.
26. JCS 2124/11, 27 July 1950, and JCS 2073/115, 27 February 1951.
27. JCS 2073/61, 3 September 1950, and JCS 2073/140, 27 March 1951.
28. Poole, *op. cit.*, p. 189, and JCS 2073/71, 26 September 1950.
29. Poole, *op. cit.*, pp. 170–171.

30. JCS 2143/6, "Joint Outline War Plan for A War Beginning 1 July 1954." Reaper (Reaper was later renamed *Groundwork* and then *Headstone*), 29 November 1950.
31. *Ibid.*, V Summary of Strategic Estimate.
32. *Ibid.*, VII Overall Strategic Concept.
33. *Ibid.*, VIII Basic U.S. Undertakings.
34. SAC Emergency War Plan approved by JCS on 22 October 1951. The plan was also intended to support Reaper. See Poole, *op. cit.*, pp. 169–170.
35. JCS 2143/6, *op. cit.*, XIII First Phase Tasks.
36. *Ibid.*
37. *Ibid.*
38. *Ibid.*, XII Phased Concept of Operations.
39. *Ibid.*
40. *Ibid.*, Appendixes A, B, and C.
41. *Ibid.*, Appendixes D, E, and F.
42. *Ibid.*, XII.
43. JCS 2143/24, 4 September 1952.
44. *Ibid.*, and Poole, *op. cit.*, pp. 172–174.
45. Poole, *op. cit.*, pp. 175–176.
46. JCS 2073/179, 30 July 1951.
47. JSPC 876/349, 9 August 1951.
48. JCS 2073/210, 12 September 1951.
49. JLPC 416/112, 24 August 1951.
50. JCS 2073/228, 26 October 1951 (approved 30 October).
51. JCS 1844/116, 2 November 1951.
52. *Ibid.*
53. JLPC 416/136, 2 December 1951.
54. JIC 558/80, 28 December 1951.
55. JSPC 876/414, 29 December 1951.
56. JCS 2220/2, 18 January 1952, and JCS 2220/2, 23 January 1952.
57. JIC 636/2, 11 June 1953.
58. JIC 636/2, 16 June 1953.
59. JCS 1924/75, 20 October 1953; JCS 1924/75, 21 October 1953; and JCS 1924/76, 29 October 1953.

CONCLUSION

From late 1945 until the early 1950s JCS war, mobilization, and logistics plans had a number of common characteristics. American planners assumed that the USSR did not intend to launch a general war. Rather, they believed that Moscow sought to create a Soviet-dominated Communist world order by a combination of subversion and intimidation. War, if it came, would come not by design but by accident. The Russians might misjudge the willingness of a Western nation to resist pressure; troops in contact in occupied areas might clash, or the success of Western European recovery might frighten Moscow into a premature attack.

No matter how the war began it would be started by the Russians. Every single American plan assumed that the war would open with a series of Soviet offensives. Even after the Soviet Union amassed an atomic arsenal, JCS planners never seriously contemplated a first strike. They recognized the advantages of being the first to launch an atomic offensive and even contemplated a launch-on-warning response but finally accepted the fact that in no case would the United States preempt the USSR.

The JCS assumed that a conflict with the USSR would be global and unlimited. In the absence of political guidance from the civilian leadership, military planners had to establish their own national policy objectives. Rejecting unconditional surrender, American plans, nevertheless, demanded a price so high that only Russian capitulation would permit the United States to achieve its political goals. NSC20 finally provided a definitive statement of national objectives. Like the military plans, NSC20 did not seek unconditional surrender, but the

151

government's objectives were so extensive that attaining them required a situation that resembled unconditional surrender in everything but name.

Thus, any conflict with the USSR would be fought to the finish. None of the plans even mentioned a surrogate war, and none contemplated a limited war. Nor did the plans deal with the prospect of war termination short of complete victory. Even the Korean War did not change this basic outlook. Although the United States fought a limited conflict and engaged in protracted negotiations while battles continued, the JCS maintained the conviction that a war between the U.S. and the USSR would resemble the course and outcome of World War II.

The grand strategy of World War III would also follow the pattern of the Second World War. The conflict would begin with massive successful enemy offensives. Europe and the Middle East would be the major theaters with Asian battle zones playing important but distinctly secondary roles. The Allies—the United States, United Kingdom, most Commonwealth members, and a number of lesser powers—would mobilize their resources and halt the enemy drives. Airpower would cripple the enemy industrial base, and the Allies would mount large-scale counteroffensives. Thus, World War III, like its two predecessors, would not only be total but also protracted. Most plans presumed a conflict lasting at least three years.

The Chiefs of Staff viewed the special relationship between Washington and London as a cornerstone of American strategy. For several years after 1945 the military planners regarded Great Britain as America's only reliable great power ally, and even after the formation of NATO, the Anglo-American relationship was viewed as critically important. The preservation of the United Kingdom despite occasional doubts about Great Britain's ability to survive an initial Soviet onslaught almost always found its place as a basic undertaking of U.S. war plans.

The JCS focused on Western Eurasia as the decisive region in a future conflict, but operational concepts underwent significant changes. Initially, the American military believed that Western Europe was indefensible. The Russians were simply too strong for the Allies to do more than offer token resistance on the Continent. Early plans, therefore, called for a rapid retreat from Europe and contained no concept of a second Normandy. Against the might of the Red Army there was little or no prospect of success by direct assault.

The Allies would instead build up their forces in the Middle East, especially in the Cairo-Suez area. From Cairo-Suez the Allies could mount in conjunction with forces operating from England an air offensive against Soviet industry and transportation. A bastion in the Middle East would also allow the Allies to protect the region's oil resources, and finally, the Allies would mount their counterattack from Middle Eastern bases into the Black Sea and the Ukraine.

European recovery and the creation of the WEU and, later, NATO led American strategists to change their priorities. Planners recognized the political and strategic importance of Western Europe and the necessity of reassuring the European allies of America's willingness to participate directly and actively in their defense. Concepts of operations evolved to include a fighting retreat from the Continent. The next phase was to establish a Continental bridgehead and if this proved infeasible to organize a reinvasion as quickly as possible. By the early 1950s, emergency plans contemplated holding a line on the Pyrenees while mid-range plans called for a forward defense which at a minimum held a line from Zeeland to the Rhine and from there to the Alps and the Piave River.

The role of the Middle East remained ambivalent. It was always viewed as an important base region, especially for strategic bombers. Some plans claimed that Middle Eastern oil was vital to the Allied war effort and that the oil-producing regions either had to be held or, if lost, recovered as soon as possible. Other plans, however, had no qualms about relinquishing the oil regions and assumed that the Allies would either sabotage or destroy by bombing oil wells and production facilities. By the start of the 1950s the JCS tended to view the Middle East as an important region but secondary in priority to the defense of Europe.

The importance and impact of the atomic bomb was also unclear. Several early plans did not include provisions for an A-bomb offensive since the military felt that atomic weapons might be outlawed or placed under international control. On the other hand, some strategists believed that the atomic bomb was a vital new weapon, and others, especially from the Air Force, felt that the A-bomb would be in and of itself decisive in any future war. Of course, nobody knew for sure how many bombs it would take to destroy the USSR's ability to sustain a war. Moreover, there were many other unanswered questions about atomic warfare: Were there enough atomic weapons? Was there sufficient accurate target data? Could the Air Force sustain an air offensive at an acceptable loss rate? Would an atomic offensive, even if most weapons struck their intended targets, be effective militarily and psychologically?

The Harmon and WSEG reports indicated that atomic weapons could indeed inflict substantial damage upon the USSR. The damage would not, however, be sufficient to drive Russia out of the war, and aircraft loss rates would be very high. By 1949–1950 an emerging concensus seemed to indicate that atomic weapons were *a* decisive weapon but *not* the decisive weapon in a general war. A-bombs would thus play a role similar to that of air power in World War II. In conjunction with conventional land and sea forces air power, especially atomic air power, could produce victory.

Atomic weapons became a keystone of American strategy not only because of their inherent power but also because post-war fiscal austerity left the American military with inadequate conventional forces. American strategic thinkers could devise no method to match Soviet conventional forces other than by relying heavily on atomic munitions. By 1947–1948 plans Broiler and Frolic involved little more than having the Army protect forward air bases, the Navy hold the lines of communication to the bases and the Air Force as soon as possible after D-Day launch an atomic offensive.

The creation of the WEU and NATO forced a strategic reevaluation, but even plans that committed the United States to an active defense of Western Europe continued to rely heavily on atomic weapons. In the absence of sufficient conventional armaments the JCS felt it had no choice but to place atomic warfare in the forefront of American strategy.

The advent of a Soviet atomic arsenal had little impact on American war plans other than additional demands for more atomic weapons to hit a new set of targets that included Russian atom bomb production and storage facilities and airfields for long-range bombers. Planners recognized that the Russians would ultimately attain the ability to cripple the US industrial base, but this realization produced no new strategic initiatives. Strategists continued to reject preemption and sought to defend vital targets while waiting for D-Day to launch forces against Russian atomic and industrial capabilities.

The JCS regarded the atomic bomb primarily as a more effective iron bomb. Atomic missions resembled the kind of strikes flown in World War II. Plans never mentioned the concept of deterrence, and when the military did use the term, they used it in the traditional sense of America's overall military and economic might. Atomic weapons were then treated as more powerful, more effective conventional munitions rather than as weapons that were qualitatively different from all previous instruments of war.

Whether or not the JCS plans would have been followed and would in case of war have led the United States to victory is, of course, conjectural. The USSR might indeed have collapsed under SAC's atomic offensive. On the other hand, the Red Army might have overrun Continental Europe, neutralized the British Isles, taken the Middle East, and though hurt by atom bomb attacks forced American power back into the Atlantic. Washington would then have been faced with the grim prospect of fighting alone against a Eurasian giant.

Fortunately, the JCS plans never had to be tested in battle. The United States government evidentally took a calculated risk by drastically reducing its armed forces after 1945 and maintaining austere defense budgets throughout the late 1940s. The risk was based upon the assumption that Moscow would not in fact resort to war. Insufficient conventional forces deprived the American military of much operational flexibility and virtually compelled the JCS to rely heavily upon atomic weapons, but in a broader sense the risk paid off. No war between the United States and the Soviet Union took place and despite the vast military and political changes that have occurred since the 1950s this basic fact so far remains unchanged.

BIBLIOGRAPHY

Primary Sources - JCS Documents

JCS Info Memo 374, 5 Feb 1945.

JCS 1545, 9 October 1945.

JCS 203rd Meeting, 16 October 1945.

JIS 80/7, 23 October 1945.

JIS 80/10, 25 October 1945.

JIS 80/9, 26 October 1945.

JCS 1477/1, Overall Effect of Atomic Bomb on Warfare and Military Organization, 30 October 1945

JIS 80/11, 31 October 1945.

JIS 80/12, 1 November 1945.

JLPC 35/11, November 1945.

JIC 329, Strategic Vulnerability of the USSR to a Limited Air Attack, 3 November 1945.

JIS 80/14, Estimate of Soviet Postwar Military Capabilities and Intentions, 8 November 1945.

JIC 332, 16 November 1945.

JIC 250/6, 29 November 1945.

JIC 329/1, Strategic Vulnerability of the USSR to a Limited Air Attack, 3 December 1945.

SWNCC 282, 19 December 1945.

JWPC Committee 416/1, Military Position of the United States in the Light of Russian Policy, 8 January 1946.

JCS 1477/5, 12 January 1946.

JIC 341, Aims and Sequence of Soviet Political and Military Moves, 31 January 1946.

JIC 342, 6 February 1946.

JIS 226/2, Areas Vital to the Soviet War Effort, 12 February 1946.

JIS 226/3, Areas Vital to the Soviet War Effort, 4 March 1946.

JPS 789, Concept of Operations for "Pincher," 2 March 1946.

JCS 1641, Request by Secretary of State for Strategic Estimate of Security Interests in Eastern Mediterranean, 7 March 1946.

JCS 1545/1, Military Position of the United States in Light of Russian Policy, 9 March 1946.

JCS 1641/1, 10 March 1946.

JIC 342/2, 27 March 1946.

JCS 1477/10, 31 March 1946.

JCS 1641/4, 6 April 1946.

JCS 1641/5, Estimate Based on Assumption of Occurrence of Major Hostilities, 11 April 1946.

JPS 789/1, Staff Studies of Certain Military Problems Deriving from Concept of Operations for Pincher, 13 April 1946.

JWPC 432/3, Joint Basic Outline War Plan Short Title "Pincher," 27 April 1946.

JLPC, 21 May 1946.

SWNCC 38/35, June 5, 1946.

JSP Tentative Over-All Strategic Concept and Estimate of Initial Operations Short Title "Pincher," 14 June 1946.

JWPC 432/7, Tentative Over-All Strategic Concept and Estimate of Initial Operations Short Title "Pincher," 18 June 1946.

JLPC 35/16, 27 June 1946.

JIS 80/26, 9 July 1946.

JIS 253/1, 26 July 1946.

JCS, Clark Clifford to Leahy, 18 July Response, 26 July 1946.

JCS Decision Amending JCS 1696, 27 July 1946.

JWPC 458/1, Preparation of Joint Plan Broadview, August 5, 1946.

JCS 467/1, Griddle, 15 August 1946.

JIS 249/3, August 23, 1946.

JPS 815, October 24, 1946.

JWPC 475/1, Caldron, 2 November 1946, Defense of area Italy-India.

JIC 374/1, Intelligence Estimate Assuming that War Between Soviet and Non-Soviet Powers Breaks Out in 1956, 6 November 1946.

JCS 1725, 6 November 1946.

JLPC 35/27, 21 November 1946.
JIC 375/1, 29 November 1946.
JWPC 464/1, "Cockspur," 2 December 1946.
JIC 374/2, 8 January 1947.
JIS 267/1, 14 January 1947.
JIC 237, 17 January 1947.
JCS 1725/1, 13 February 1947, 30 April 1947.
JCS 1745/1, 25 February 1947.
JIS 274/1, Soviet Logistic Capabilities for Support of Iberian Campaign and Air Assault on Great Britain, 5 March 1947.
JIS 274/1, 26 March 1947.
JIS 275/1, 18 April 1947.
JWPC 474/1, 15 May 1947.
JCS Final Report of the Joint Chiefs of Staff Evaluation Board for Operation Crossroads, June 30, 1947.
JWPC 846/7, 29 July 1947.
JWPC 465/2, "Drumbeat," 4 August 1947.
JWPC 486/8, Guidance for Mobilization Planning as Affected by Loss of the Mediterranean Line of Communications, 18 August 1947.
JWPC 476/2, "Moonrise," 29 August 1947.
JLPC 17/11, 2 September 1947.
SWNCC 38/46, 8 September 1947.
JCS 1805, 23 September 1947.
JWPC 473/1, "Deerland," 30 September 1947.
JCS 1770/1, 7 October 1947.
JCS 1796/1, 14 October 1947.
JSPG 496/1, "Broiler," 8 September 1947.
JCS 1725/13, 14 November 1947.
JLPC 416/8, 3 December 1947.
JLPC 416/9, 3 December 1947.
JSPC 846/7, Aircraft Requirements, 3 December 1947.
JSPG 499/2, "Charioteer," 3 December 1947.
JCS 1745/5, 8 December 1947.
JCS 1796/6, Aircraft Requirements, 10 December 1947.
JCS 1725/14, 23 January 1948.
JSPG 496/4, "Broiler," 12 February 1948.
JIC 380/2, Soviet threat '48–52, 16 February 1948.
Munitions Board Report, 18 February 1948.
JCS 1837, 20 February 1948.

JCS 1844, Emergency Plans, 9 March 1948.

JSPG 500/2, "Bushwacker," 8 March 1948.

JCS 1884/1, Short Range Emergency Plan Grabber (Originally *Frolic* changed to *Grabber*, 5 April 1949), 17 March 1948.

JLPG. Quick Feasibility Test of JSPG 496/4, 19 March 1948.

JIG 286/1, Intelligence Estimate on Espionage Subversion and Sabotage, 22 March 1948.

SANACC 176/38, 22 March 1948.

JIC Policy memo no. 2, 29 March 1948.

JCS 1844/2, 6 April 1948.

JCS 1483/54, 9 April 1948.

JCS 1828/1, 17 April 1948.

JCS 1868/1, 17 April 1948.

JSPC, Implementation of "Frolic," 19 April 1948.

JCS 1860/5, 20 April 1948.

JSPC 876, 21 April 1948.

JSPC 877/2, Future Planning, 22 April 1948.

JSPC, 27 April 1948.

JCS 1844/3, 27 April 1948.

JCS 1770, 28 April 1947.

JCS 1805/6, 28 April 1948.

JSPC 877/3, 3 May 1948.

JSPG 496/11, Directives for the Implementation of "Frolic," 4 May 1948.

SANACC 398/1, "Preparations for Demolition of Oil Facilities in the Middle East," 6 May 1948.

JSPG 496/10, "Crankshaft," 11 May 1948.

JSPC 877/6, 16 May 1948.

JCS 1844/4, 19 May 1948.

JCS 893/5, 24 May 1948.

SANACC 398/4, 25 May 1948.

JCS 1844/7, Directives for the Implementation of Doublestar (formerly Fleetwood and Halfmoon), 26 May 1948.

JLPC 416/12, Logistic Feasibility of Operations Planned, "Halfmoon," 15 June 1948.

JCS 1805/7, 15 June 1948.

JCS 1725/19, 17 June 1948.

JCS 1844/9, Brief of an Alternative Short Range Emergency War Plan Grabber (Frolic), 18 June 1948.

JLPC 416/13, The Joint Logistics Plan for Emergency Plan "Halfmoon," 24 June 1948.

JLPC 416/4, The Joint Logistic Plan for Emergency Plan "Frolic" (Grabber), 25 June 1948.

JSPC 877/9, 21 July 1948.

JCS 1844/13, Directives for the Implementation of Doublestar (Fleetwood - Halfmoon), 22 July 1948.

JCS 1745/15, A-bomb Assembly Teams, 27 July 1948.

JCS 1844/14, 11 August 1948.

JCS 1844/15, The Logistic Feasibility of Doublestar (Fleetwood - Halfmoon), 12 August 1948.

JCS 1844/16, Directive for Development and Coordination of Logistic Plans for Doublestar (Halfmoon - Fleetwood), 13 August 1948.

JCS 1920, 16 August 1948.

JIC 429/1, 25 August 1948.

JCS 1725/22, "Cogwheel," 26 August 1948.

JCS 1725/23, 27 August 1948.

JCS 1844/19, Directive for Implementation of Doublestar, 1 September 1948.

JCS 1844/20, Brief of an Alternative Short-Range Emergency War Plan, 3 September 1948.

JCS 1725/24, 10 September 1948.

JCS 1725/25, 14 September 1948.

JIG 286/2, 17 September 1948.

JIC (48) 76 (0) Final, Fighting Values Russian-Allied Forces, 21 September 1948.

JCS 1844/26, 22 September 1948.

JCS 1868/22, 10 October 1948.

JSPC 877/23, Revised Brief of Short-Range Emergency Plan Short Title: Fleetwood (Halfmoon - Doublestar), 14 October 1948.

JCS 1725/27, 19 October 1948.

JLPC 416/32, The Logistic Feasibility of ABC 101, 12 November 1948.

JCS 1844/31, Coordinated Emergency War Plans, Allied Occupation Forces Austria and Trieste, 23 November 1948.

JIC 435/12, 30 November 1948.

JIG 278/6, 23 December 1948.

JCS 1844/33, 6 January 1949.

JCS 1844/34, 18 January 1949.

JCS 1920/1, Report by the *ad hoc* Committee Long-Range Plans for War with the USSR - Development of a Joint Outline Plan for Use in the Event of War in 1957, 31 January 1949.

JCS 1725/34, 18 February 1949.

JCS 1725/35, 25 February 1949.

JCS 1725/36, 28 February 1949.

JCS 1920/2, 7 March 1949.

JCS 1868/63, Outline of Short-Term Plan (1 July 1949) for Defense of Western Europe, 8 March 1949.

JCS 1920/3, 18 March 1949.

JCS 1920/4, 4 April 1949.

Metric, 26 April 1949.

JCS 1725/42, 27 April 1949.

JCS 1844/37, Preparation of a Joint Outline Emergency War Plan, 27 April 1949.

JCS 1844/39, 13 May 1949.

JCS 1823/14, 27 May 1949.

JCS 1868/86, 28 May 1949.

JCS 1844/41, 18 June 1949.

Metric, 6 July 1949.

JCS 1725/45, 9 July 1949.

JCS 1844/38, 14 July 1949.

JCS 1868/16, 5 August 1949.

JCS 1725/48, 9 August 1949.

JCS 1868/99, 16 August 1949.

JCS 1924/6, 3 September 1949.

JSPC 876/38, 7 September 1949.

JCS 1868/136, 25 October 1949.

JCS 1725/64, 8 November 1949.

JCS 1844/46, Joint Outline Emergency War Plan Offtackle (Shakedown - Crosspiece), 8 November 1949.

JCS 1844/47, Logistic Implications of Offtackle, 15 November 1949.

JCS 2084, 16 November 1949.

JCS 1868/149, 22 November 1949.

JCS 1725/67, 1 December 1949.

JCS 1844/49, 7 December 1949.

JCS 1920/5 (Vol. I), Long-Range Plans for War with the USSR— Development of a Joint Outline Plan for use in the Event of War in 1957 (Short Title: *Dropshot*), 19 December 1949.

JCS 1920/5 (Vol. II), 19 December 1949.

JCS 1920/5 (Vol. III), 19 December 1949.

JCS 1805/18, 9 January 1950.

JIC 500, 13 January 1950.

JCS 1844/53, 26 January 1950.

JCS 2084/2, 31 January 1950.

Joint Advanced Study Committee Pattern of War in the Atomic Warfare Age, 15 February 1950.

JCS 1844/55, Directives for Implementation of Offtackle, 18 February 1950.

JCS 2073/6, 20 February 1950.

JCS 2073/7, 27 February 1950.

JCS 2086/1, 5 April 1950.

JCS 2128, 4 May 1950.

JSPC 876/124, 18 May 1950.

JCS 2084/9, 11 May 1950.

JSPC 876/130, 25 May 1950.

Munitions Board Report, 25 May 1950.

JCS 1776/6, 29 June 1950.

JCS 1924/10, 30 June 1950.

JCS 1776/13, 2 July 1950.

JCS 2073/41, 12 July 1950.

JCS 1923/28, 14 July 1950.

JCS 2124/11, 27 July 1950.

JCS 1823/29, 3 August 1950.

JIC 530/3, 22 August 1950.

JCS 2073/61, 3 September 1950.

JCS 2073/72, 26 September 1950.

JCS 1776/122, 3 October 1950.

JCS 2143/6, Joint Outline War Plan for a War beginning 1 July 1954 "Reaper," 29 November 1950.

JCS 2118/4, 27 December 1950.

JCS 2118/10, 12 January 1951.

JCS 1924/49, Estimate of the Scale and Nature of the Immediate Communist Threat to the Security of the United States, 5 February 1951.

JIC 435/52, 7 February 1951.

JCS 1837/18, 12 February 1951.

JSPC 958/15, 16 February 1951.

JCS 2073/115, 25 February 1951.

JCS 2073/140, 27 March 1951.

JCS 1924/53, 29 March 1951.

JCS 2118/19, 2 April 1951.

JCS 1924/54, 3 April 1951.

JCS 1924/57, Estimate of Sino-Soviet Capabilities in the Far East with Respect to Japan, 17 May 1951.

JCS 1924/59, 26 June 1951.

JCS 1924/60, Soviet Capabilities and Intentions in the Far East, 10 July 1951.

JIC 554/15, Estimate of Communist Preparations for Offensive Operations in the Far East, 23 July 1951.

JCS 2073/179, 30 July 1951.

JIC 558/33, 8 August 1951.

JLPC 416/112, 24 August 1951.

JCS 2073/228, 26 October 1951.

JCS 1844/116, Guidance for Emergency War Plan (Majestic 1844/116), 2 November 1951.

JSPC 887/7, 30 November 1951.

JIC 558/80, 28 December 1951.

JSPC 876/414, 29 December 1951.

JCS 2220/2, 18 January 1952.

JCS 220/3, 23 January 1952.

JCS 1924/65, 11 April 1952.

JSPC 958/46, 18 April 1952.

JCS 1924/67, 24 July 1952.

JCS 1924/68, 7 August 1952.

JCS 2143/24, 4 September 1952.

JCS 2143/25, 30 October 1952.

JCS 1924/69, 25 November 1952.

JLPC 416/136, 9 December 1952.

JCS 1844/137, 29 January 1953.

JCS 1844/138, 7 February 1953.

JCS 1924/70, 19 February 1953.

JCS 2220/19, 6 May 1953.

JIC 636/2, 11 June 1953.

JIC 636/2, 16 June 1953.

JCS 1924/75, 20 October 1953.

JCS 1924/75, 21 October 1953.

JCS 1924/76, 29 October 1953.

Secondary Works

Albion, R.G., and R.H. Connery. *Forrestal and the Navy*. New York: Columbia University Press, 1962.

Ambrose, Stephen. *Eisenhower, Volume One. Soldier General of the Army President-Elect 1890–1952*. New York: Simon and Schuster, 1983.

Borklund, C.W. *The Department of Defense*. New York: Frederick A. Praeger, 1968.

Borowski, Harry R. *A Hollow Threat: Strategic Air Power and Containment before Korea*. Westport, Conn.: Greenwood Press, 1982.

Bradley, Omar R., and Clay Blair. *A General's Life*. New York: Simon and Schuster, 1983.

Caraley, D. *The Politics of Military Unification, A Study of Conflict and the Policy Process*. New York: Columbia University Press, 1966.

Cave Brown, Anthony. *Drop Shot, The United States Plan for War with the Soviet Union in 1957*. New York: Dial Press, 1978.

Coletta, P.E. "The Defense Unification Battle, 1947–50: The Navy," in *Prologue*, Vol. 7, No. 1. Spring 1975.

Condit, Kenneth W. *The History of the Joint Chiefs of Staff, The Joint Chiefs of Staff and National Policy. Volume II. 1947–1949*. Wilmington, Del.: Michael Glazier, Inc., 1979.

Davis, B.V. *Admirals, Politics and Post War Defense Policy: The Origins of the Postwar U.S. Navy, 1943–1946 and After*. Ph.D Dissertation. Princeton University , 1962.

Donnelly, Charles H. *United States Defense Policies since World War II*. Washington, D.C.: GPO, 1957.

Etzold, Thomas H., and John L. Gaddis, eds. *Containment: Documents on American Policy and Strategy 1945–1950*. New York: Columbia University Press, 1978.

Evangelista, Matthew A. "Stalin's Postwar Army Reappraised." *International Security*, Vol. 7, No. 3. Winter 1982/1983.

First Report of the Secretary of Defense, 1948. Washington, D.C.: GPO, 1949.

Freedman, Lawrence. *The Evolution of Nuclear Strategy*. New York: St. Martin's Press, 1983.

Goldberg, Alfred, ed. *A History of the United States Air Force 1907–1957*. Princeton, N.J.: D. Van Nostrand Co., 1957.

Herken, Gregg. *The Winning Weapon: The Atomic Bomb in the Cold War 1945–1950*. New York: Alfred A. Knopf, 1980.

Kinnard, Douglas. *The Secretary of Defense*. Lexington: University of Kentucky Press, 1980.

Korb, Lawrence. *The Joint Chiefs of Staff. The First Twenty-five Years*. Bloomington: Indiana University Press, 1976.

Legere, L., Major. *Unification of the Armed Forces*. Cambridge, Mass.: Harvard University Archives, 1950.

Love, Robert William, Jr., ed. *The Chiefs of Naval Operations*. Annapolis, Md.: Naval Institute Press, 1980.

Matloff, Maurice. *American Military History*. Washington, D.C.: OCMH, 1973.

Millis, Walter. *Arms and Men: A Study in American Military History*. New York: G.P. Putnam's Sons, 1956.

Poole, Walter S. *The History of the Joint Chiefs of Staff, The Joint Chiefs of Staff and National Policy, Vol. IV, 1950–1952*. Wilmington, Del.: Michael Glazier, 1980.

Radford, Arthur, and Stephen Jurika, eds. *From Pearl Harbor to Vietnam*. Stanford, Calif.: Hoover Institution Press, 1980.

Rearden, Steven L. *History of the Office of the Secretary of Defense, Volume I, The Formative Years*. Washington, D.C.: Historical Office OSD, 1984.

Rosenberg, David A. "American Atomic Strategy and the Hydrogen Bomb Decision." *Journal of American History*, Vol. 66, No. 1, June 1979.

————. "U.S. Nuclear Stockpile 1945 to 1950." *Bulletin of the Atomic Scientists*, Vol. 38, No. 5, May 1982.

————. "The Origins of Overkill: Nuclear Weapons and American Strategy, 1945–1960." *International Security*, Vol. 7, No. 4, Spring 1983.

Schnabel, James F. *The History of the Joint Chiefs of Staff: The Joint Chiefs of Staff and National Policy, Volume I, 1945–1947.* Wilmington, Del., 1979.

Second Report of the Secretary of Defense, 1949, Washington, D.C.: GPO, 1949.

Semi-Annual Report of the Secretary of Defense, 1949, Washington, D.C.: GPO, 1950.

Weigley, Russell F. *History of the United States Army.* New York: Macmillan Publishing Co., 1967.

Additional Works

President's Air Policy Commission. *Survival in the Air Age Report.* Washington, D.C.: GPO, 1948.

US Department of the Air Force. *Report of the Secretary of the Air Force for Fiscal Year 1948.* Washington, D.C.: GPO, 1949.

US Department of the Army. *Report of the Secretary of the Army for Fiscal Year 1948.* Washington, D.C.: GPO, 1949.

US Department of the Navy. *Annual Report of the Secretary of the Navy for the Fiscal Year 1948.* Washington, D.C.: GPO, 1949.

US Department of State. *Documents on Disarmament, 1945–1949.* 2 vols. Washington, D.C.: GPO, 1960.

————. *Foreign Relations of the United States, 1946.* Vol. 1: *General; The United Nations.* Washington, D.C.: GPO, 1972.

————. *1946.* Vol. 6: *Eastern Europe; The Soviet Union.* Washington, D.C.: GPO, 1969.

————. *1947.* Vol. 1: *General; The United States Nations.* Washington, D.C.: GPO, 1973.

————. *1949.* Vol. 1: *National Security Affairs; Foreign Economic Policy.* Washington, D.C.: GPO, 1976.

————. *1950.* Vol. 1: *National Security Affairs; Foreign Economic Policy.* Washington, D.C.: GPO, 1977.

————. *1950.* Vol. 3: *Western Europe.* Washington, D.C.: GPO, 1976.

————. *The International Control of Atomic Energy: Growth of a Policy.* Washington, D.C.: GPO, 1946.

————. *1950.* Vol. 3: *Korea.* Washington, D.C.: GPO, 1946.

————. *A Short History of the US Department of State, 1781–1981.* Publication 9166. Washington, D.C.: US Department of State, January 1981.

US Department of State. Committee on Atomic Energy. *A Report on the International Control of Atomic Energy.* Prepared for the Secretary of State's Committee on Atomic Energy by a Board of Consultants. Washington, D.C.: GPO, 1946.

US National Archives and Records Service. *Public Papers of the Presidents: Harry S. Truman.* 8 vols., 1945–53. Washington, D.C.: GPO, 1961–66.

US National Military Establishment, Office of the Secretary of Defense. *First Report of the Secretary of Defense.* Washington, D.C.: GPO, 1948.

US Department of Defense, Office of the Secretary of Defense, Annual and Semi-Annual Reports, Fiscal Year 1949 through Fiscal Year 1950. Washington, D.C.: GPO, 1949–51.

US Office of Civil Defense Planning. *Civil Defense for National Security, Report to the Secretary of Defense*. Washington, D.C.: GPO, 1948.

US Strategic Bombing survey. *The Effects of Atomic Bombs on Hiroshima and Nagasaki*. Pacific Survey Report No. 3. Washington, D.C.: GPO, 1946.

———. *Japan's Struggle to End the War*. Washington, D.C.: GPO, 1946.

———. *The Strategic Air Operations of Very Heavy Bombers in the War against Japan (Twentieth Air Force)*. Washington, D.C.: GPO, 1946.

US President's Advisory Commission on Universal Training. *A Program for National Security, Report*, May 29, 1947. Washington, D.C.: GPO, 1947.

Congressional Documents

The basic Congressional Document source for the evolution of United States defense policy since World War II are the four sets of Congressional hearings produced annually during the defense authorization and appropriations process since Fiscal Year 1949. These include authorization hearings in the House and Senate Armed Services committees, and appropriations hearings in the defense subcommittees of the House and Senate Appropriations committees. In order to conserve space, these hearings are not listed separately below. Other hearings and reports with particular relevance to the evolution of nuclear strategy are as follows:

US Congress, House. *United States Defense Policies since World War II*. House document, No. 100, by C.H. Donnelly, 85/1. Washington, D.C.: GPO, 1957.

———. Committee on Armed Services. *Investigation of the B-36 Bomber Program, Hearings*, 81/1. Washington, D.C.: GPO, 1949.

———. *The National Defense Program: Unification and Strategy, Hearings*, 81/1. Washington, D.C.: GPO, 1950.

———. *Investigation of the B-36 Bomber Program, Report No. 1470*, 81/2. Washington, D.C.: GPO, 1950.

———. *Unification and Strategy, Document No. 60*, 81/2. Washington, D.C.: GPO, 1950.

————. *To Convert the National Military Establishment into an Executive Department of Defense, and to Provide the Secretary of Defense with Appropriate Responsibility and Authority*, 81/2. Washington, D.C.: GPO, 1949.

————. Committee on Military Affairs. *An Act for the Development and Control of Atomic Energy, Hearings*, 79/2. Washington, D.C.: GPO, 1945.

————. *House Concurrent Resolution 80, Composition of the Postwar Navy, Hearings*, 79/2. Washington, D.C.: GPO, 1945.

————. Joint Committee on the Investigation of the Pearl Harbor Attack. *Investigation of the Pearl Harbor Attack, Report*, 79/2. Washington, D.C.: GPO, 1946.

US Congress, Joint Congressional Aviation Policy Board. *National Aviation Policy, Report No. 949*, 80/2. Washington, D.C.: GPO, 1948.

————. Senate, Committee on Armed Services. *National Defense Establishment (Unification), Hearings on S. 758*, Parts 1–3, 80/1. Washington, D.C.: GPO, 1947.

————. *Universal Military Training, Hearings*, 80/2. Washington, D.C.: GPO, 1948.

————. *National Security Act Amendments of 1949, Hearings*. Washington, D.C.: GPO, 1949.

————. Committee on Armed Services and Committee on Foreign Relations. *Assignment of Ground Forces of the United States to Duty in the European Area, Hearings*, 82/1. Washington, D.C.: GPO, 1951.

————. *Military Situation in the Far East, Hearings*, Parts 1–5, 82/1. Washington, D.C.: GPO, 1951.

————. Vol. 12: *86th Congress, 2nd Session, 1960*. Washington, D.C.: GPO, 1982.

————. Committee on Military Affairs. *Department of Armed Forces, Hearings*, on S. 84 and S. 1482, 79/1. Washington, D.C.: GPO, 1945.

————. Committeee on Naval Affairs. *Unification of the War and Navy Departments and Postwar Organization for National Security*, 79/1. Washington, D.C.: GPO, 1945.

————. Special Committee on Atomic Energy. *Atomic Energy Act of 1946, Hearings*, 79/1–79/2. Washington, D.C.: GPO, 1943–46.

Limited Circulation Official Government or Government Contractor
Studies and Histories

Air Defense Command, Headquarters, Directorate of Historical Services. *The Air Defense of the United States through June 1951*. Colorado Springs, Colo.: Air Defense Command, 1952.

————. *Emergency Air Defense Forces, 1946–1954*. Historical Study 5. Colorado Springs, Colo.: Air Defense Command, 1954. ˙

Alexander, Frederic C., Jr. *History of Sandia Corporation through Fiscal Year 1963*. Albuquerque, N.M.: The Sandia Corporation, 1963.

Bowen, Lee. *History of the Air Force Atomic Energy Program*. Vol. IV: *The Development of Weapons*. Washington, D.C.: US Air Force Historical Division, 1955. Declassified with deletions, June 1981.

Brodie, Bernard, and Galloway, Eilene. *The Atomic Bomb and the Armed Services*. Public Affairs Bulletin No. 55. Washington, D.C.: Library of Congress Legislative Reference Service, May 1947.

Cooling, Benjamin Franklin. *The Army Support of Civil Defense, 1945–1966: Plans and Policy*. Monograph No. 108M. Washington, D.C.: US Department of the Army, Office of the Chief of Military History, 1969.

Davis, Vernon E. *The History of the Joint Chiefs of Staff in World War II, Organizational Development*. Vol. 1: *Origin of the Joint and Combined Chiefs of Staff*. Vol. 2: *Development of the JCS Committee Structure*. Washington, D.C.: Joint Chiefs of Staff, Historical Division of the Joint Secretariat, 1972.

Doughty, Major Robert A. *The Evolution of US Army Tactical Doctrine, 1946–76*. Fort Leavenworth, Kan.: Combat Studies Institute, Leavenworth Papers, August 1979.

Futrell, Robert Frank. *Ideas, Concepts, Doctrine: A History of Basic Thinking in the US Air Force, 1907–1964*. Maxwell Air Force Base, Ala.: Aerospace Studies Institute, 1971.

Goldberg, Alfred. *A Brief Survey of the Evolution of Ideas about Counterforce*, RM-5431-PR. Santa Monica, Calif.: The Rand Corporation, October 1967; revised, March 1981.

Goldberg, Alfred (ed.). *History of Headquarters, USAF, 1 July 1949 to 30 June 1950*; and *History of Headquarters, USAF, 1 July 1950 to 30 June 1951*. Washington, D.C.: Department of the Air Force, 1955. Declassified with deletions, 1975.

Greer, Thomas H. *Development of Air Doctrine in the Army Air Arm, 1917–1941*. Maxwell Air Force Base, Ala.: Air University, 1953.

James, Staff Sergeant Martin E. *Historical Highlights, United States Air Forces in Europe, 1945–1979*. Office of History, US Air Forces in Europe, November 1980.

Joint Chiefs of Staff, Joint Secretariat, Historical Division. *Major Changes in the Organization of the Joint Chiefs of Staff, 1942–1977*. Washington, D.C.: Joint Chiefs to Staff, 1978.

Lemmer, George F. *The Air Force and the Concept of Deterrence, 1945–1950*. Washington, D.C.: USAF Historical Division Liaison Office, 1963. Declassified with deletions, 1975.

———. *The Air Force and Strategic Deterrence, 1951–1960*. Washington, D.C.: Office of Air Force History, 1967. Declassified with deletions, 1980.

Little, Robert D. *The History of Air Force Participation in the Atomic Energy Program, 1943–1953*. Vol. 2: *Foundations of an Atomic Air Force and Operation SANDSTONE, 1946–1948*. Washington, D.C.: Air University Historical Liaison Office, 1955. Declassified with deletions, 1981.

———. *Organizing for Strategic Planning, 1945–1950: The National System and the Air Force*. Washington, D.C.: USAF Historical Division Liaison Office, April 1964. Declassified with deletions, 1975.

McClandon, R. Earl. *Unification of the Armed Forces. Administrative and Legislative Development, 1945–1949*. Maxwell Air Force Base, Ala.: Air University, April 1952.

McMullen, Richard F. *Air Defense and National Policy, 1946–1950*. Historical Study 22. Colorado Springs, Colo.: Air Defense Command, 1964.

Naval Research Advisory Committee. *Report on Historical Perspectives in Long Range Planning in the Navy*. Washington, D.C.: Assistant Secretary of the Navy (Research Engineering and Systems), 1981.

Ponturo, John. *Analytical Support for the Joint Chiefs of Staff: The WSEG Experience, 1948–1976*, Study S-507. Arlington, Va.: Institute for Defense Analyses, July 1979.

Rosenberg, David Alan, and Kennedy, Floyd D. *History of the US – USSR Strategic Arms Competition, Supporting Study: US Aircraft Carriers in the Strategic Role, Part I: Naval Strategy in a Period of Change: Strategic Interaction, Interservice Rivalry, and the Development of a Nuclear Attack Capability, 1945–1951*. Deputy Chief of Naval Operations (Plans and Policy) Contract No. N00014-75-C00237. Falls Church, Va.: Lulejian and Associates, October 1975.

Rosenburg, Max. *The Air Force and the National Guided Missile Program 1944–1950.* Washington, D.C.: USAF Historical Division Liaison Office, June 1964. Declassified 1981.

Smith, Richard K. *Cold War Navy.* Chief of Information, US Navy, Contract No. N00014-75-C-1001. Falls Church, Va.: Lulejian and Associates, March 1976.

Sparrow, John C. *History of Personnel Demobilization in the United States Army,* Pamphlet No. 20-210. Washington, D.C.: Department of the Army, July 1952.

Strategic Air Command, Headquarters, Office of the Historian. *Development of Strategic Air Command, 1946–1981.* Omaha, Nebr.: Office of the Historian, Headquarters, Strategic Air Command, 1982.

———. *History of Strategic Air Command,* 1946. Offutt Air Force Base, Nebr.: Headquarters, Strategic Air Command, 1950.

———. 1947. Offutt Air Force Base, Nebr.: Headquarters, Strategic Air Command, 1949.

———. 1948. Offutt Air Force Base, Nebr.: Headquarters, SAC, 1949.

———. 1949. Offutt Air Force Base, Nebr.: Headquarters, SAC, 1950.

———. January–June 1950. Offutt AFB, Nebr.: Headquarters, SAC, 1950.

Strategic Air Command, Headquarters, Office of the Historian. *History of Strategic Air Command,* July–December 1950. Offutt Air Force Base, Nebr.: Headquarters, Strategic Air Command, 1951.

———. January–June 1951. Offutt AFB, Nebr.: Headquarters, SAC, 1951.

———. Headquarters, Historical Division. *SAC Targeting Concepts,* Historical Study 73A. Omaha, Nebr.: Strategic Air Command, 1959. Declassified with deletions, 1980, 1981.

US Department of Energy, Office of Public Affairs, Nevada Operations Office. *Announced United States Nuclear Tests, July 1945 through December 1981,* NVO-209 (Rev. 2), January 1982.

US Navy, Office of the Chief of Naval Operations. *US Naval Aviation in the Pacific: A Critical Review.* Washington, D.C.: Office of the Chief of Naval Operations, 1947.

Books and Scholarly Articles

Acheson, Dean. *Present at the Creation. My Years in the State Department.* New York: W.W. Norton, 1969.

Albion, Robert G., and Connery, Robert H. *Forrestal and the Navy.* New York: Columbia University Press, 1962.

Albion, Robert G., *Makers of Naval Policy, 1798–1947*. Annapolis, Naval Institute Press 1980.

Alpedrovitz, Gar. *Atomic Diplomacy: Hiroshima and Potsdam*. New York: Simon and Schuster, 1965.

Anders, Roger (ed.). *Forging the Atomic Shield: The Atomic Energy Commission Diaries of Gordon A. Dean*. Chapel Hill, N.C.: University of North Carolina Press, 1987.

Andrews, Marshall. *Disaster Through Air Power*. New York: Rhinehart, 1950.

Arkin, William M.; Cochran, Thomas B.; and Hoenig, Milton M. "The US Nuclear Stockpile." *Arms Control Today* 12 (April 1982) 1–2.

Arneson, R. Gordon. "The H-Bomb Decision." *Foreign Service Journal* 46 (May 1969) 27–29.

Arnold, Henry H. *Global Mission*. New York: Harper and Brothers, 1949.

Aron, Raymond. *The Great Debate. The Theories of Nuclear Strategy*. Translated by Ernst Pawel. Garden City, N.Y.: Anchor Books, 1965.

Bartlett, C.J. *The Long Retreat. A Short History of British Defence Policy*. London: St. Martin's Press, n.d.

Batchelder, Robert C. *The Irreversible Decision, 1939–1950*. New York: Macmillan, 1961; Macmillan Paperbacks Edition, 1965.

Beard, Edmund. *Developing the ICBM: A Study in Bureaucratic Politics*. New York: Columbia University Press, 1976.

Bernstein, Barton J. (ed.). *The Atomic Bomb. The Critical Issues*. Boston: Little, Brown, 1976.

Bernstein, Barton J. "The Challenges and Dangers of Nuclear Weapons: American Foreign Policy and Strategy, 1945–1961." *Foreign Service Journal* 55 (September 1978) 9–15, 36.

Bernstein, Barton J. "Doomsday II." *New York Times Magazine*. July 27, 1975.

———. "The Perils and Politics of Surrender: Ending the War with Japan and Avoiding the Third Atomic Bomb." *Pacific Historical Review*, 44, No. 1 (February 1977) 1–27.

———. "The Quest for Security: American Foreign Policy and International Control of the Atomic Bomb, 1942–1946." *The Journal of American History* 60 (March 1974) 1003–1044.

———. "Roosevelt, Truman, and the Atomic Bomb, 1941–1945: A Reinterpretation." *Political Science Quarterly* 90 (Spring 1975) 23–69.

Betts, Richard K. *Soldiers, Statesmen, and Cold War Crises*. Cambridge: Harvard University Press, 1977.

Blackett, P.M.S. *Fear, War, and the Bomb. Military and Political Consequences of Atomic Energy.* New York: McGraw Hill, 1949.

————. *Studies of War, Nuclear and Conventional.* New York: Hill and Wang, 1962.

Blum, John Morton (ed.). *The Price of Vision. The Diary of Henry A. Wallace, 1942–1946.* Boston: Houghton Mifflin, 1973.

Blumberg, Stanley A., and Owens, Gwinn. *Energy and Conflict. The Life and Times of Edward Teller.* New York: G.P. Putnam's Sons, 1976.

Bollinger, Lynn L.; Lilley, Tom; and Lombard, Albert E. "Preserving American Air Power", Harvard Business Review 23 (Spring 1945) 372–92.

Borden, William Liscom. *There Will Be No Time. The Revolution In Strategy.* New York: Macmillan, 1946.

Borklund, Carl W. *Men of the Pentagon. From Forrestal to McNamara.* New York: Frederick A. Praeger, 1966.

Borowski, Harry R. "Air Force Atomic Capability from V-J Day to the Berlin Blockade – Potential or Real?" *Military Affairs* 44 (October 1980) 105–110.

Borowski, Harry. *A Hollow Threat. Containment and Strategic Air Power before Korea.* Westport, Conn.: Greenwood Press, 1982.

Bottome, Edgar M. *The Balance of Terror. A Guide to the Arms Race.* Boston: Beacon Press, 1971.

Bradley, Omar N. and Blair, Clay. *A General's Life: An Autobiography.* N.Y.: Simon and Schuster, 1983.

Brodie, Bernard (ed.). *The Absolute Weapon. Atomic Power and World Order.* New York: Harcourt, Brace, 1946.

Brodie, Bernard, and Galloway, Eilene. *The Atomic Bomb and the Armed Services.* Washington, D.C.: Legislative Reference Service, Library of Congress, May 1947.

Brodie, Bernard. "New Tactics in Naval Warfare." *Foreign Affairs* 24 (January 1946) 210–233.

————. *Strategy in the Missile Age.* Princeton, N.J.: Princeton University Press, 1959.

————. *War and Politics.* New York: Macmillan, 1973.

Brown, Seyom. *The Faces of Power. Constancy and Change in United States Foreign Policy from Truman to Johnson.* New York: Columbia University Press, 1968.

Brown, Anthony Cave (ed.). *Dropshot. The United States Plan for War with the Soviet Union, 1957.* New York: Dial Press, 1978.

Brown, Anthony Cave, and MacDonald, Charles B. (eds.). *The Secret History of the Atomic Bomb*. New York: Dell Publishing Company, 1977.

Buell, Thomas B. *Master of Sea Power. A Biography of Fleet Admiral Ernest J. King*. Boston: Little, Brown, 1980.

Bush, Vannevar. *Modern Arms and Free Men. A Discussion of the Role of Science in Preserving Democracy*. New York: Simon and Schuster, 1949.

———. *Pieces of the Action*. New York: William Morrow, 1970.

Byrnes, James F. *All in One Life Time*. New York: Harper and Brothers, 1958.

———. *Speaking Frankly*. New York: Harper and Brothers, 1947.

Campbell, Thomas M., and Herring, George C. (eds.). *The Diaries of Edward R. Stettinius, Jr., 1943–1946*. New York: New Viewpoints.

Caraley, Demetrios. *The Politics of Military Unification. A Study of Conflict and the Policy Process*. New York: Colombia University Press, 1966.

Cate, James L. "Development of Air Doctrine, 1917–1941." *Air University Review* 1 (Winter 1947) 11–22.

Clausewitz, Carl von. *On War*. Translated and edited by Michael Howard and Peter Paret. Princeton, N.J.: Princeton University Press, 1976.

Cline, Ray S. *Washington Command Post. The Operations Division*. Washington, D.C.: GPO, 1950.

Coffey, Thomas M. *Hap. The Story of the US Air Force and the Man Who Built It, General Henry H. "Hap" Arnold*. New York: The Viking Press, 1962.

Collins, J. Lawton. *Lightning Joe. An Autobiography*. Baton Rouge, La.: Louisiana State University Press, 1979.

Cole, Alice C.; Goldberg, Alfred; Tucker, Samuel A.; and Winnacker, Rudolph. *The Department of Defense. Documents on Establishment and Organization*. Washington, D.C.: GPO, 1978.

Coles, Harry L. (ed.). *Total War and Cold War. Problems in Civilian Control of the Military*. Columbus, Ohio: Ohio State University Press, 1962.

Coletta, Paolo E. *The United States Navy and Defense Unification, 1947–1953*. Newark, Del.: University of Delaware Press, 1981.

Coletta, Paolo E. *American Secretaries of the Navy*. 2 Vols. Annapolis Naval Institute Press, 1980.

Conant, James B. *My Several Lives. Memoirs of a Social Inventor*. New York: Harper and Row, 1970.

Cutler, Robert. *No Time for Rest*. Boston: Atlantic – Little Brown and Company, 1956.

Davis, Vincent. *The Admirals' Lobby.* Chapel Hill, N.C.: University of North Carolina Press, 1967.

———. *The Politics of Innovation. Patterns in Navy Cases.* Denver, Colo.: University of Denver, 1967.

———. *Postwar Defense Policy and the US Navy, 1943–1946.* Chapel Hill, N.C.: University of North Carolina Press, 1966.

De Santis, Hugh. *The Diplomacy of Silence. The American Foreign Service, the Soviet Union, and the Cold War, 1933–1947.* Chicago: University of Chicago Press, 1980.

Dickens, Admiral Sir Gerald. *Bombing and Strategy. The Fallacy of Total War.* London: Sampson Low, Marston, 1947.

Dingman, Roger "Strategic Planning and the Policy Process: American Plans for War in East Asia, 1945–1950." *Naval War College Review* 32 (November–December 1979): 4–21.

Dingman, Roger "Atomic Diplomacy During the Korean War." *International Security* Vol. 13, No. 3, (Winter 1988–89) 50–91.

Donovan, Robert J. *Conflict and Crisis. The Presidency of Harry S. Truman, 1945–1948.* New York: W.W. Norton, 1977.

———. *Tumultuous Years. The Presidency of Harry S. Truman, 1949–1953.* New York: W.W. Norton, 1982.

Douhet, Giulio. *The Command of the Air.* New York: Coward and McCann, 1942.

Eastman, James N., Jr. "Flight of the Lucky Lady II." *Aerospace Historian* 16 (Winter 1969): 9–11, 33–5.

Earle, Edward Mead. "The Influence of Air Power Upon History." *The Yale Review* 35 (June 1946): 577–93.

Egan, Clifford L., and Knott, Alexander W. (eds.). *Essays in Twentieth Century Diplomatic History Dedicated to Professor Daniel M. Smith.* Washington, D.C.: University Press of America, 1982.

Emme, Eugene M. (ed.). *The History of Rocket Technology.* Detroit, Mich.: Wayne State University Press, 1964.

Emme, Eugene M. (ed.). *The Impact of Air Power: National Security and World Politics*, New York, Van Nostrand, 1959.

———. *The Impact of Air Power.* New York: D. Van Nostrand, 1959.

Emmerson, John K. *The Japanese Thread. A Life in the US Foreign Service.* New York: Holt, Rinehart and Winston, 1978.

Etzold, Thomas H., and Gaddis, John Lewis (eds.). *Containment: Documents on American Policy and Strategy, 1945–1950.* New York: Columbia University Press, 1978.

Feis, Herbert. *The Atomic Bomb and the End of World War II*. Princeton, N.J.: Princeton University Press, 1961; Princeton Paperback, 1970.

———. *From Trust to Terror. The Onset of the Cold War, 1945–1950*. New York: Norton, 1970.

Ferrell, Robert H. *The American Secretaries of State and Their Diplomacy*. Vol. 15: New York: Cooper Square, 1966.

Ferrell, Robert H. (ed.). *The Eisenhower Diaries*. New York: W.W. Norton, 1981.

———. *Off the Record. The Private Papers of Harry S. Truman*. New York: Harper and Row, 1980.

Finletter, Thomas K. *Power and Policy. US Foreign Policy and Military Power in the Hydrogen Age*. New York: Harcourt, Brace, 1954.

Food, Rosemary. *The Wrong War: American Policy and the Dimensions of the Korean War*. Ithaca, N.Y.: Cornell University Press, 1985.

Freedman, Lawrence. *The Evolution of Nuclear Strategy*. New York: St. Martin's, 1981.

———. *US Intelligence and the Soviet Strategic Threat*. Boulder, Colo.: Westview Press, 1977.

Friedberg, Aaron L. "A History of the U.S. Strategic 'Doctrine' – 1945 to 1980." *The Journal of Strategic Studies* (3 December 1980) 37–71.

Futrell, Robert Frank. *The United States Air Force in Korea, 1950–1953*. New York, Duell, Sloan and Pearce, 1961.

Gaddis, John Lewis. "Reconsiderations. Was the Truman Doctrine a Real Turning Point?" *Foreign Affairs* 5 (January 1974) 386–402.

Gaddis, John Lewis. *Strategies of Containment. A Critical Appraisal of Postwar American National Security Policy*. New York: Oxford University Press, 1982.

———. *The United States and the Origins of the Cold War, 1941–1947*. Columbia University Press, 1972.

Gallery, Daniel V. *Eight Bells and All's Well*. New York: W.W. Norton, 1965.

George, Alexander L., and Smoke, Richard. *Deterrence in American Foreign Policy. Theory and Practice*. New York: Columbia University Press, 1974.

Gerber, Larry G. "The Baruch Plan and the Origins of the Cold War." *Diplomatic History* 6 (Winter 1982): 69–95.

Gertsch, W. Darrell. "The Strategic Air Offensive and the Mutation of American Values, 1937–1945." *Rocky Mountain Social Science Journal* 2 (October 1974) 37–50.

Gilpin, Robert. *American Scientists and Nuclear Weapons Policy.* Princeton, N.J.: Princeton University Press, 1962.

Glazier, Kenneth M., Jr. "The Decision to Use Atomic Weapons against Hiroshima and Nagasaki." *Public Policy* 18 (1969) 463–516.

Goldberg, Alfred (ed.). *A History of the United States Air Force, 1907–1957.* Princeton, J.J.: D. Van Nostrand, 1957.

Goulden, Joseph C. *Korea. The Untold Story of the War.* New York: Times Books, 1982.

Gowing, Margaret. *Independence and Deterrence. Britain and Atomic Energy, 1945–1952.* Vol. 1: *Policy Making.* Vol. 2: *Policy Execution.* London: St. Martin's, (1974).

Graybar, Lloyd J. "Bikini Revisited." *Military Affairs* 44 (October 1980) 118–123.

Greene, Fred. "The Military View of American National Policy, 1904–1940." *The American Historical Review* 66 (January 1961) 354–77.

Greenfield, Kent Roberts (ed.). *Command Decisions.* Washington, D.C.: GPO. 1960.

Groueff, Stephane. *Manhattan Project. The Untold Story of the Making of the Atomic Bomb.* Boston: Little Brown, 1967.

Groves, Leslie R. *Now It Can Be Told. The Story of the Manhattan Project.* New York: Harper, 1962.

Gunston, Bill. *Bombers of the West.* New York: Charles Scribner's Sons, 1973.

Hagan, Kenneth J., (ed.). In Peace and War: *Interpretations of American Naval History 1775–1978.* Westport, Greenwood, 1978.

Hammond, Thomas T. "Atomic Diplomacy Revisited." *Orbis* 19 (Winter 1976) 1403–1428.

Hammond, Paul Y. *Organizing for Defense. The American Military Establishment in the Twentieth Century.* Princeton, N.J.: Princeton University Press, 1961.

Hansell, Haywood J. *The Air Plan that Defeated Hitler.* Atlanta, Ga.: Higgins, McArthur, 1972.

Hansell, Major General H.S. "Strategic Air Warfare." *Aerospace Historian* 13 (Winter 1966) 153–160.

Hansen, Charles. "Nuclear Neptunes." *Journal of the American Aviation Historical Society* 24 (Winter 1979) 262–68.

Hansen, Charles. "US Nuclear Bombs." *Replica in Scale* 3 (January 1976) 154–59.

Hayes, Grace P. *The History of the Joint Chiefs of Staff in World War II: The War against Japan.* Annapolis, MD.: US Naval Institute Press, 1982.

Haynes, Richard F. *The Awsome Power. Harry S. Truman as Commander in Chief.* Baton Rouge, La.: Louisiana State University Press, 1973.

Heller, Francis H. (ed.). *The Truman White House. The Administration of the Presidency, 1945–1953.* Lawrence, Kans.: Regents Press of Kansas, 1980.

Henrikson, Alan K. "The Map as an Idea: the Role of Cartographic Imagery during the Second World War." *The American Cartographer* 2 (1975) 19–53.

Herken, Gregory Franklin. "A Most Deadly Illusion: The Atomic Secret and American Nuclear Weapons Policy, 1945–1950." *Pacific Historical Review* 49 (February 1980) 51–76.

———. *The Winning Weapon: The Atomic Bomb in the Cold War, 1945–1950.* New York: Aldred A. Knopf, 1980.

Herring, George C., Jr. *Aid to Russia, 1941–1946. Strategy, Diplomacy, and the Origins of the Cold War.* New York: Colombia University Press, 1973.

Hewes, James. *From Root to McNamara: Army Organization and Administration 1900–1963.* Washington, D.C., Center of Military History, 1975.

Hewlett, Richard G., and Anderson, Oscar E. *A History of the United States Atomic Energy Commission.* Vol. 1: *The New World, 1939–1946.* University Park, Penna.: Pennsylvania State University Press, 1962.

Hewlett, Richard G., and Duncan, Francis. *History of the United States Atomic Energy Commission.* Vol. 2: Atomic Shield, 1947–1952. University Park, Penna.: Pennsylvania State University Press, 1969.

———. *Nuclear Navy, 1946–1962.* Chicago: University of Chicago Press, 1974.

Higham, Robin. *Air Power: A Concise History.* New York, St. Martins, 1973.

Holley, I.B., Jr. *An Enduring Challenge. The Problem of Air Force Doctrine.* Harmon Memorial Lectures in Military History No. 16. Colorado Springs, Colo.: US Air Force Academy, 1974.

Holloway, David. "Research Note: Soviet Thermonuclear Development." *International Security* 4 (Winter 1979–80): 192–97.

Hubler, Richard G. *SAC. The Strategic Air Command.* New York: Duell, Sloan, and Pierce, 1958.

Huie, William Bradford. *The Case against the Admirals. Why We Must Have a Unified Command.* New York: E.P. Dutton, 1946.

Huntington, Samuel P. *The Common Defense. Strategic Programs in National Politics.* New York: Columbia University Press, 1961.

Hurley, Alfred F., and Ehrhardt, Robert (eds.). *Air Power and Warfare. The Proceedings of the 8th Military History Symposium, US Air Force Academy, October 18–20, 1978.* Washington, D.C., GPO, 1979.

Ireland, Timothy P. *Creating the Entangling Alliance. The Origins of the North Atlantic Treaty Organization.* Westport, Conn.: Greenwood Press, 1981.

Jackson, Henry M. (ed.). *The National Security Council: Jackson Subcommittee Papers on Policy-Making at the Presidential Level.* New York: Praeger, 1965.

Jungk, Robert. *Brighter than a Thousand Suns. A Personal History of the Atomic Scientists.* Translated by James Cleugh. New York: Harcourt, Brace, 1958.

Jurika, Stephen, Jr. (ed.). *From Pearl Harbor to Vietnam. The Memoirs of Admiral Arthur W. Radford.* Stanford, Calif.: Hoover Institution Press, 1980.

Karman, Theodore von, with Lee Edson. *The Wind and Beyond. Theodore von Karman. Pioneer in Aviation and Pathfinder in Space.* Boston: Little, Brown, 1967.

Karten, David. "The Bomber that Never Bombed." *Aerospace Historian* 13 (Winter 1966) 161–63.

Kaufmann, William W. (ed.). *Military Policy and National Security.* Princeton, N.J.: Princeton University Press, 1956.

Kennan, George F. *Memoirs.* Vol. 1: *1925–1950.* Vol 2: *1950–1963.* Boston: Little, Brown, 1969, 1972.

Kinnard, Douglas. *The Secretary of Defense.* Lexington, Ky.: University Press of Kentucky, 1980.

Kolko, Joyce, and Kolko, Gabriel. *The Limits of Power: The World and United States Foreign Policy, 1945–1954.* New York: Harper and Row, 1972.

Kolodzej, Edward A. *The Uncommon Defense and Congress, 1945–1963.* Columbus, Ohio: Ohio State University Press, 1966.

Koppes, Clayton R. *JPL and the American Space Program. A History of the Jet Propulsion Laboratory.* New Haven, Conn: Yale University Press, 1982.

Kramish, Arnold. *Atomic Energy and the Soviet Union.* Stanford, Calif.: Stanford University Press, 1959.

Kunetka, Jams W. *City of Fire. Los Alamos and the Birth of the Atomic Age, 1943–1945.* Englewood Cliffs, N.J.: Prentice-Hall, 1978.

Lamont, Lansing. *Day of Trinity.* New York: Atheneum, 1965.

Lapp, Ralph E. *Must We Hide?* Cambridge, Mass.: Addison-Wesley Press, 1949.

Leahy, William D. *I Was There.* New York: McGraw Hill, 1950.

LeMay, General Curtis E., with MacKinley Kantor. *Mission with LeMay. My Story.* Garden City, N.Y.: Doubleday, 1965.

Liddell Hart, B.H. *The Revolution in Warfare.* New Haven, Conn.: Yale University Press, 1947.

Lieberman, Joseph. *The Scorpion and the Tarantula.* Boston: Houghton, Mifflin, 1970.

Lilienthal, David E. *The Journals of David E. Lilienthal.* Vol. 2: *The Atomic Energy Years, 1945–1950.* New York: Harper and Row, 1964.

Love, Robert William, Jr. (ed.). *The Chiefs of Naval Operations.* Annapolis, Md.: US Naval Institute Press, 1980.

MacCloskey, Monro. *The United States Air Force.* New York, Praeger, 1967.

Maddox, Robert J. *The New Left and the Origins of the Cold War.* Princeton, N.J.: Princeton University Press, 1973.

Mandelbaum, Michael. *The Nuclear Question. The United States and Nuclear Weapons, 1946–1976.* London and New York: Cambridge University Press, 1979.

Marshall, George C.; Arnold, H.H.; and King, Ernest J. *The War Reports of General of the Army George C. Marshall, General of the Army H.H. Arnold, and Fleet Admiral Ernest J. King.* Philadelphia: J.B. Lippincott, 1947.

Matloff, Maurice, and Snell, Edwin S. *Strategic Planning for Coalition Warfare, 1941–1942.* Washington, D.C.: GPO, 1953.

Matloff, Maurice. *Strategic Planning for Coalition Warfare, 1943–1944.* Washington, D.C.: GPO, 1959.

May, Ernest R. "Development of Political-Military Consultation in the United States." *Political Science Quarterly* 70 (June 1955): 161–80.

May, Ernest R. (ed.). *The Ultimate Decision. The President as Commander in Chief.* New York: George Brazillier, 1960.

Mazusan, George T. "America's U.N. Commitment 1945–1953." *The Historian* 40 (February 1978) 309–30.

McLellan, David S. *Dean Acheson. The State Department Years.* New York: Dodd, Mead, 1976.

McNeill, William Hardy. *America, Britain, and Russia. Their Cooperation and Conflict, 1941–1946.* New York: Johnson Reprint, 1970.

————. *The Pursuit of Power. Technology, Armed Force, and Society since A.D. 1000*. Chicago: University of Chicago Press, 1982.

Melosi, Martin V. *The Shadow of Pearl Harbor. Political Controversy over the Surprise Attack, 1941–1946*. College Station, Texas: Texas A & M University Press, 1977.

Messer, Robert L. *The End of an Alliance. James F. Byrnes, Roosevelt, Truman, and the Origins of the Cold War*. Chapel Hill, N.C.: University of North Carolina Press, 1982.

Millett, Allan R. *Semper Fidelis. The History of the United States Marine Corps*. New York: Macmillan, 1980.

Millett, Allan R. and Maslowski, Peter. *For the Common Defense: A Military History of the United States of America*. New York, The Free Press, 1984.

Millis, Walter. *Arms and the State. Civil-Military Elements in National Policy*. New York: Twentieth Century Fund, 1958.

Millis, Walter, and Duffield, Eugene (eds.). *The Forrestal Diaries*. New York: Viking, 1951.

Morison, Elting E. *Turmoil and Tradition. A Study of the Life and Times of Henry L. Stimson*. Boston: Houghton Mifflin, 1960.

Morse, Philip M. *In at the Beginnings. A Physicist's Life*. Cambridge, Mass.: MIT Press, 1977.

Morton, Louis. "The Decision to Use the Atomic Bomb." *Foreign Affairs* 25 (January 1957) 334–53.

————. "War Plan Orange, Evolution of a Strategy." *World Politics* 11 (January 1959) 221–50.

Moss, Norman. *Men Who Play God. The Story of the Hydrogen Bomb*. New York: Harper and Row, 1969; Penguin Books Reprint with revisions, 1972.

Mrozek, Donald J. "A New Look at 'Balanced Forces': Defense Continuities from Truman to Eisenhower." *Military Affairs* 38 (December 1974) 145–151.

Newman, James R., and Miller, Byron S. *The Control of Atomic Energy. A Study of its Social, Economic, and Political Implications*. New York: McGraw Hill, 1948.

Nichols, K.D. *The Road to Trinity* (N.Y.: William Morrow, 1987).

O'Connor, Raymond G. *Diplomacy for Victory. FDR and Unconditional Surrender*. New York: W.W. Norton, 1971.

Osgood, Robert Endicoot. *NATO. The Entangling Alliance*. Chicago: University of Chicago Press, 1962.

Pierre, Andrew J. *Nuclear Politics. The British Experience with an Independent Strategic Force, 1939–1970.* London: Oxford University Press, 1972.

Polmar, Norman. *Aircraft Carriers. A Graphic History of Carrier Aviation and Its Influence on World Events.* Garden City, N.Y.: Doubleday, 1969.

———. *Strategic Weapons. An Introduction.* Revised. New York: Cran, Russak, 1982.

Poole, Walter S. "From Conciliation to Containment. The Joint Chiefs of Staff and the Coming of the Cold War, 1945–1946." *Military Affairs* 42 (February 1978) 12–16.

Possony, Stefan T. *Strategic Air Power. The Pattern of Dynamic Security.* Washington, D.C.: Infantry Journal Press, 1949.

Potter, E.B. *Nimitz.* Annapolis, MD.: Naval Institute Press, 1976.

Prados, John. *The Soviet Estimate. US Intelligence Analysis and Russian Military Strength.* New York: Dial Press, 1982.

Pringle, Peter, and Spiegelman, James. *The Nuclear Barons.* New York: Holt, Rinehart and Winston, 1981.

Quester, George H. *Nuclear Diplomacy. The First Twenty-Five Years.* New York: Dunellen, 1970.

Reinhardt, Colonel G.C., and Kintner, Lieutenant Colonel W.R. *Atomic Weapons in Land Combat.* Harrisburg Penns.: The Military Service Publishing Company, 1953.

Reynolds, Clark G. *The Fast Carriers. The Forging of an Air Navy.* New York: McGraw Hill, 1968.

Richardson, Brig. General Robert C., III. "NATO Nuclear Strategy: A Look Back." *Strategic Review* 9 (Spring 1981) 35–43.

Ridgway, Matthew W., with Harold H. Martin. *Soldier.* New York: Harper and Row, 1956.

Ries, John C. *The Management of Defense. Organization and Control of the US Armed Forces.* Baltimore, Md.: Johns Hopkins University Press, 1964.

Roberts, Chalmers M. *The Nuclear Years. The Arms Race and Arms Control, 1945–1970.* New York: Praeger, 1970.

Rogow, Arnold A. *James Forrestal. A Study of Personality, Politics, and Policy.* New York: Macmillan, 1963.

Rose, John P. *the Evolution of US Army Nuclear Doctrine, 1945–1980.* Boulder, Colo.: Westview Press, 1980.

Rose, Lisle. *After Yalta.* New York: Scribner, 1973.

Rosenberg, David Alan. "American Atomic Strategy and the Hydrogen Bomb Decision." *The Journal of American History* 66 (June 1979) 62–87.

———. "The Origins of Overkill: Nuclear Weapons and American Strategy, 1945–1960." *International Security* 7 (Spring 1983) 3–71.

———. "'A Smoking Radiating Ruin at the End of Two Hours': Documents of American Plans for Nuclear War with the Soviet Union, 1954–1955." *International Security* 6 (Winter 1981/1982) 3–38.

———. "The US Navy and the Problem of Oil in a Future War: the Outline of a Strategic Dilemma, 1945–1950." *Naval War College Review* 29 (Summer 1976) 33–64.

———. "US Nuclear Stockpile, 1945–1950." *The Bulletin of the Atomic Scientists* 38 (May 1982) 25–30.

Rositzke, Harry. *The CIA's Secret Operations.* New York: Reader's Digest Books, 1977.

Rowe, James Les. *Project W-47.* Livermore, Calif.: JA A RO Publications, 1978.

Ryan, Paul B. *First Line of Defense. US Navy since 1945.* Stanford, Calif.: Hoover Institution Press, 1981.

Sander, Aldred D. "Truman and the National Security Council: 1945–1947." *The Journal of American History* 59 (September 1972) 369–388.

Schilling, Warner R. "The H-Bomb Decision: How to Decide Without Actually Choosing." *Political Science Quarterly* 76 (March 1961) 24–46.

Schilling, Warner R.; Hammond, Paul Y.; and Snyder, Glenn H. *Strategy, Politics, and Defense Budgets.* New York: Columbia University Press, 1962.

Schoenberger, Walter Smith. *Decision of Destiny.* Athens, Ohio: Ohio University Press, 1960.

Schwarz, Jordan A. *The Speculator. Bernard Baruch in Washington, 1917–1963.* Chapel Hill, N.C.: University of North Carolina Press, 1981.

Sherry, Michael S. *Preparing for the Next War, American Plans for Post-war Defense, 1941–1945.* New Haven, Conn.: Yale University Press, 1977.

———. "The Slide to Total Air War." *The New Republic* (16 December 1981) 20–25.

Sherwin, Martin J. "The Atomic Bomb and the Origins of the Cold War: US Atomic Energy Policy and Diplomacy, 1941–45." *American Historical Review* 78 (October 1973) 945–67.

————. *A World Destroyed. The Atomic Bomb and the Grand Alliance.* New York: Knopf, 1975.

Shurcliff, W.A. *Bombs at Bikini. The Official Report of Operation Crossroads.* New York: William H. Wise, 1947.

Smith, Alice Kimball, *A Peril and a Hope. The Scientists' Movement in America, 1945–1947.* Chicago: University of Chicago Press, 1965.

Smith, Alice Kimball, and Weiner, Charles (eds.). *Robert Oppenheimer. Letters and Recollections.* Cambridge: Harvard University Press, 1960.

Smith, Dale O. *US Military Doctrine.* Boston: Little, Brown, 1955.

Smith, Gaddis. *The American Secretaries of State and Their Diplomacy.* Vol. 16: *Dean Acheson.* New York: Cooper Square, 1972.

Smith, Perry McCoy. *The Air Force Plans for Peace, 1943–1945.* Baltimore, Md.: Johns Hopkins University Press, 1970.

Smyth, Henry D. *Atomic Energy for Military Purposes. The Official Report on the Development of the Atomic Bomb under the Auspices of the United States Government, 1940–1945.* Princeton, N.J.: Princeton University Press, 1948.

Stein, Harold (ed.). *American Civil-Military Decision. A Book of Case Studies.* Tuscaloosa, Ala.: University of Alabama Press, 1963.

Stimson, Henry L. "The Decision to Use the Atomic Bomb." *Harper's* 194 (February 1947) 54–5.

Stimson, Henry L., and Bundy, McGeorge. *On Active Service in Peace and War.* New York: Harper and Brothers, 1947.

Stratemeyer, George E. "Administrative History of US Army Air Forces." *Air Affairs* 1 (Summer 1974) 510–25.

Strauss, Lewis L. *Men and Decisions.* Garden City, N.Y.: Doubleday, 1962.

Sturm, Thomas A. "American Air Defense: the Decision to Proceed." *Aerospace Historian* 19 (December 1972) 188–194.

Taylor, Maxwell D. *Swords and Plowshares.* New York: W.W. Norton, 1972.

Teller, Edward, and Brown, Allen. *The Legacy of Hiroshima.* Garden City, N.Y.: Doubleday, 1962.

Tibbetts, Paul W., with Claire Stebbins and Harry Frank. *The Tibbetts Story.* New York: Stein and Day, 1978.

Trachtenberg, Marc, "A 'Wasting Asset': American Strategy and the Shifting Nuclear Balance, 1949–1954" *International Security*, Vol. 13. No. 3, Winter 1988–89, 5–49.

Truman, Harry S,. *Memoirs by Harry S. Truman.* Vol. 1: *Year of Decisions,* Vol. 2: *Years of Trial and Hope, 1946–1952.* Garden City, N.Y.: Doubleday, 1955, 1956.

Twining, Nathan F. *Neither Liberty nor Safety.* New York: Holt, Rinehart, and Winston, 1966.

Ulam, S.M. *Adventures of a Mathematician.* New York: Charles Scribner's Sons, 1976.

US Atomic Energy Commission. *In the Matter of J. Robert Oppenheimer: Transcript of Hearings before Personnel Security Board, and Texts of Principal Documents and Letters.* Cambridge, Mass.: MIT Press, 1971.

US Joint Chiefs of Staff. *The History of the Joint Chiefs of Staff. The Joint Chiefs of Staff and National Policy:*

Schnabel, James F. *The History of the Joint Chiefs of Staff. The Joint Chiefs of Staff and National Policy.* Vol. 1: *1945–1947.* Wilmington, Del.: Michael Glazier, 1979.

Condit, Kenneth W. *The History of the Joint Chiefs of Staff. The Joint Chiefs of Staff and National Policy.* Vol. 2: *1947–1949.* Wilmington, Del.: Michael Glazier, 1979.

Schnabel, James F., and Watson, Robert J. *The History of the Joint Chiefs of Staff. The Joint Chiefs of Staff and National Policy.* Vol. 3: *The Korean War* 2 parts. Wilmington, Del.: Michael Glazer, 1979.

Poole, Walter S. *The History of the Joint Chiefs of Staff. The Joint Chiefs of Staff and National Policy.* Vol. 4: *1950–1952.* Wilmington, Del.: Michael Glazier, 1980.

US Navy, Naval Weapons Center. *History of the Naval Weapons Center, China Lake, California:*

Christman, Albert B. *History of the Naval Weapons Center, China Lake, California.* Vol. 1: *Sailors, Scientists, and Rockets.* Washington, D.C.: GPO, 1971.

Gerrard-Gough, J.D., and Christman, Albert B. *History of the Naval Weapons Center, China Lake, California.* Vol. 2: *The Grand Experiment at Inyokern.* Washington, D.C.: GPO, 1978.

Vandenberg, Arthur H., Jr. (ed.). *The Private Papers of Senator Vandenberg.* Boston: Houghton Mifflin, 1952.

Weigley, Russell F. *The American Way of War. A History of United States Military Strategy and Policy.* New York: Macmillan, 1973.

———. *History of the United States Army.* New York: Macmillan, 1967.

———. *Towards an American Army. Military Thought from Washington to Marshall.* New York: Columbia University Press, 1962.

Wells, Samuel F., Jr. "The Origins of Massive Retaliation." *Political Science Quarterly* 96 (Spring 1981) 31–52.

Wilson, Donald. "Origin of a Theory for Air Strategy." *Aerospace Historian* 18 (March 1971) 19–25.

Wolk, Herman S. "The B-29, the A-Bomb, and the Japanese Surrender." *Air Force Magazine* (February 1975) 55–61.

————. "The Defense Unification Battle, 1947–1950: The Air Force." *Prologue* 7 (Spring 1975) 18–26.

————. "Men Who Made the Air Force." *Air University Review* 23 (September–October 1972) 9–23

————. "Roots of Strategic Deterrence." *Aerospace Historian* 19 (September 1971) 137–144.

Yavenditti, Michael J. "The American People and the Use of Atomic Bombs against Japan: The 1940s" *The Historian* 36 (February 1974) 224–247.

————. "John Hersey and the American Conscience: The Reception of 'Hiroshima'." *Pacific Historical Review* 43 (February 1974) 24–49.

Yergin, Daniel. *Shattered Peace. The Origins of the Cold War and the National Security State*. Boston: Houghton Mifflin, 1977.

York, Herbert. *The Advisors. Oppenheimer, Teller, and the Superbomb*. San Francisco: W.H. Freeman, 1976.

Unpublished Papers and Dissertations

Bruins, Berend D. "Navy Bombardment Missiles to 1960." Ph.D. dissertation, Columbia University, 1981.

Dur, Philip A. "The Sixth Fleet: A Case Study of Institutional Naval Presence." Ph.D. dissertation, Harvard University, 1975.

Green, Murray. "Stuart Symington and the B-36." Ph.D. dissertation, The American University, 1960.

Herken, Gregory Franklin. "American Diplomacy and the Atomic Bomb, 1945–1947." Ph.D. dissertation, Princeton University, 1974.

Hill, William Steinert, Jr. "The Business Community and National Defense: Corporate Leaders and the Military, 1943–1950." Ph.D. dissertation, Stanford University, 1980.

Klotz, Frank G. "The President and the Control of Strategic Nuclear Weapons." Ph.D. dissertation, Oxford University, 1980.

MacIsaac, David. "The Air Force and Strategic Thought, 1945–1951." Unpublished Working Paper Number 8, International Security Studies Program, Woodrow Wilson International Center for Scholars, June 21, 1979.

Mrozek, Donald John. "Peace through Strength: Strategic Air Power and the Mobilization of the United States for the Pursuit of Foreign Policy, 1945–1955." Ph.D. dissertation, Rutgers University, 1972.

O'Brien, Larry Dean. "National Security and the New Warfare: Defense Policy, War Planning, and Nuclear Weapons, 1945–1950." Ph.D. dissertation, Ohio State University, 1981.

Parrish, Noel Francis. "Behind the Sheltering Bomb: Military Indecision from Alamagordo to Korea." Ph.D. dissertation, Rice University, 1988.

Rosenberg, David Alan. "Planning for a PINCHER War: Objectives and Military Strategy in American Planning for War with the Soviet Union, 1945–1948." Presented at the Annual Meeting of the Society for Historians of American Foreign Relations, July, 1978.

Schneider, Mark Bernard. "Nuclear Weapons and American Strategy, 1945–1953." Ph.D. dissertation, University of Southern California, 1974.

Smith, Robert London. "The Influence of U.S.A.F. Chief of Staff General Hoyt S. Vandenberg on United States National Security Policy." Ph.D. dissertation, The American University, 1965.

Wilson, Donald Edward. "The History of President Truman's Air Policy Commission and Its Influence on Air Policy, 1947–1949." Ph.D. dissertation, University of Denver, 1979.

Printed in the United States
by Baker & Taylor Publisher Services